# Nonparametric Hypothesis Testing

# Nonparametric Hypothesis Testing

## Rank and Permutation Methods
## with Applications in *R*

**Stefano Bonnini**

*University of Ferrara, Italy*

**Livio Corain**

*University of Padova, Italy*

**Marco Marozzi**

*University of Calabria, Italy*

**Luigi Salmaso**

*University of Padova, Italy*

WILEY

This edition first published 2014
© 2014 John Wiley & Sons, Ltd

*Registered office*
John Wiley & Sons Ltd, The Atrium, Southern Gate, Chichester, West Sussex, PO19 8SQ, United
Kingdom

For details of our global editorial offices, for customer services and for information about how to apply
for permission to reuse the copyright material in this book please see our website at www.wiley.com.

*Library of Congress Cataloging-in-Publication Data*

Nonparametric hypothesis testing : rank and permutation methods with applications in R / Stefano
Bonnini, Livio Corain, Marco Marozzi, Luigi Salmaso.
      pages cm
   Includes bibliographical references and index.
   ISBN 978-1-119-95237-4 (cloth)
   1. Nonparametric statistics.    2. Statistical hypothesis testing.    3. R (Computer program
language)   I. Bonnini, Stefano.    II. Corain, Livio.    III. Marozzi, Marco.    IV. Salmaso, Luigi.
   QA278.8.N64 2014
   519.5′4–dc23

                                                                                        2014020574

A catalogue record for this book is available from the British Library.

Cover image: 'A ship for discovery' by Serio Salmaso, 1980, Venice

ISBN: 978-1-119-95237-4

Set in 10/12pt Times by Aptara Inc., New Delhi, India
Printed and bound in Malaysia by Vivar Printing Sdn Bhd

1   2014

*The greatest value of a picture is when it forces us to notice what we never expected to see.*

J. Tukey

# Contents

# Presentation of the book

The importance and usefulness of nonparametric methods for testing statistical hypotheses has been growing in recent years mainly due to their flexibility, their efficiency and their ease of application to several different types of problems, including most important and frequently encountered multivariate cases. By also taking account that with respect to parametric counterparts they are much less demanding in terms of required assumptions, these peculiarities of nonparametric methods are making them quite popular and widely used even by non-statisticians.

The growing availability of adequate hardware and software tools for their practical application, and in particular of free access to software environments for statistical computing like $R$, represents one more reason for the great success of these methods.

The recognized simplicity and good power behavior of rank and permutation tests often make them preferable to the classical parametric procedures based on the assumption of normality or other distribution laws. In particular, permutation tests are generally asymptotically as powerful as their parametric counterparts in the conditions for the latter. Moreover, when data exchangeability with respect to samples is satisfied in the null hypothesis, permutation tests are always exact in the sense that their null distributions are known for any given dataset of any sample size. On the other hand, those of parametric counterparts are often known only asymptotically. Thus for most sample sizes of practical interest, the related lack of efficiency of unidimensional permutation solutions may sometimes be compensated by the lack of approximation of parametric asymptotic competitors. For multivariate cases, especially when the number of processed variables is large in comparison with sample sizes, permutation solutions in most situations are more powerful than their parametric counterparts.

For these reasons in the specialized literature a book dedicated to rank and permutation tests, problem oriented with exhaustive but simple and easy to understand theoretical explanations, a practical guide for the application of the methods to most frequently encountered scientific problems, including related $R$ codes and with many clearly discussed examples from several different disciplines, was lacking.

The present book fully satisfies these objectives and can be considered a practical and complete handbook for the application of the most important rank and permutation tests. The presentation style is simple and comprehensible also for non-statisticians with elementary education in statistical inference, but at the same time precise and formally rigorous in the theoretical explanations of the methods.

Fortunato Pesarin
*Department of Statistics*
*University of Padova*

# Preface

This book deals with nonparametric statistical solutions for hypotheses testing problems and codes for the software environment $R$ for the application of these solutions. In particular rank based and permutation procedures are presented and discussed, also considering real-world application problems related to engineering, economics, educational sciences, biology, medicine and several other scientific disciplines. Since the importance of nonparametric methods in modern statistics continues to grow, the goal of the book consists of providing effective, simple and user friendly instruments for applying these methods.

The statistical techniques are described mainly highlighting properties and applicability of the methods in relation to application problems, with the intention of providing methodological solutions to a wide range of problems. Hence this book presents a practical approach to nonparametric statistical analysis and includes comprehensive coverage of both established and recently developed methods. This 'problem oriented' approach makes the book useful also for non-statisticians. All the considered problems are real problems faced by the authors in their activities of academic counseling or found in the literature in their teaching and research activities. Sometimes data are exactly the same as in the original problem (and the data source is cited) but in most cases data are simulated and not real.

All $R$ codes are commented and made available through the book's website www.wiley.com/go/hypothesis_testing, where data used throughout the book may also be downloaded. Part of the material, including $R$ codes, presented in the book is new and part is taken from existing publications from the literature and/or from websites of different authors providing suitable $R$ codes. We fully recognize the authorship of each $R$ code and a comprehensive list of useful websites is reported in Appendix D.

The book is mainly addressed to university students, in particular for undergraduate and postgraduate studies (i.e., PhD courses, Masters, etc.), statisticians and non-statisticians experts in empirical sciences, and it can also be used by practitioners with a basic knowledge in statistics interested in the same applications described in the book or in similar problems, or consultants/experts in statistics.

Chapter 1 deals with one-sample and two-sample location problems, tests for symmetry and tests on a single distribution. First of all an introduction to rank based testing procedures and to permutation testing procedures (including nonparametric combination methodology useful for multivariate or multiple tests) is presented.

Then in this chapter, according to the number of response variables and to the number of samples, we distinguish four kinds of methods: univariate one-sample tests, multivariate one-sample tests, univariate two-sample tests and multivariate two-sample tests. In the first category the Kolmogorov–Smirnov test and the permutation test for symmetry are considered; in the second group of procedures the multivariate rank test for central tendency and the multivariate extension of the permutation test on symmetry are presented; among the procedures included in the third family of solutions the Wilcoxon test and the permutation test on central tendency are described; finally the multivariate extensions of the two-sample test on central tendency both with the rank based and permutation approach are discussed.

Chapter 2 presents some tests for comparing variabilities and distributions. For problems of variability comparisons the Ansari–Bradley test, the permutation Pan test and the permutation O'Brien test are considered. For jointly comparing central tendency and variability the Lapage test and the Cucconi test are presented. For problems related to comparisons of distributions the Kolmogorov–Smirnov and the Cramer–von Mises proposals are taken into account.

Chapter 3 is dedicated to multisample tests. For the one-way analysis of variance (ANOVA) layout the following methods are presented: the Kruskal–Wallis test, the permutation one-way ANOVA, the Mack–Wolfe test and the permutation test for umbrella alternatives. As regards the two-way ANOVA layout, the considered procedures are the Friedman test, the permutation test for related samples, the Page test for ordered alternatives and the permutation two-way ANOVA. Multiple comparison procedures for the Kruskal–Wallis test and for a permutation test are also considered. For multivariate and multisample problems a rank based and a permutation approach are presented.

Chapter 4 concerns problems for paired samples and repeated measures. For the two-sample test with paired data the Wilcoxon signed rank proposal and the permutation test for two dependent samples are discussed. For repeated measures problems the Friedman rank based test and a permutation test are considered.

Chapter 5 deals with tests for categorical data. Among one-sample problems the binomial test on one proportion, the McNemar test for paired data with binary variables and its multivariate extension are illustrated. Then two-sample tests for proportion comparisons and in general tests for $2 \times 2$ contingency tables are discussed. In particular the Fisher exact test and the permutation test for comparison of proportions are examined. The considered solutions for general problems related to $R \times C$ contingency tables are: the Anderson–Darling type permutation test, the permutation test on moments, and the chi-square permutation test.

Chapter 6 studies correlation and concordance. First, the statistical relationship between two variables is considered. The Spearman test and the Kendall test for independence are presented. Secondly, the problem of whether a set of criteria or a group of judges is concordant in ranking some objects is addressed. The Kendall–Babington Smith test and a permutation test for concordance are presented.

Finally, Chapter 7 contains a wide range of application problems and methodological solutions concerning comparisons of heterogeneity for categorical variables. The definition of statistical heterogeneity, the description of the testing problem of dominance in heterogeneity (two-sample one-sided test on heterogeneity), its two-sided and multisample extensions and the related permutation solutions are included.

We would like to express our thanks to Fortunato Pesarin, for stimulating discussions and helpful comments, to Rosa Arboretti, Eleonora Carrozzo and Iulia Cichi for helping with some *R* codes and in finding suitable reference literature. We also wish to thank Kathryn Sharples, Richard Davies and the John Wiley & Sons group in Chichester for their valuable publishing suggestions.

We welcome any suggestions for the improvement of the book and would be very pleased if the book provides users with new insights to the analysis of their data.

Stefano Bonnini
*Department of Economics and Management*
*University of Ferrara*

Marco Marozzi
*Department of Economics, Statistics and Finance*
*University of Calabria*

Livio Corain and Luigi Salmaso
*Department of Management and Engineering*
*University of Padova*

# Notation and abbreviations

ANOVA: analysis of variance

$B$: the number of conditional Monte Carlo iterations

$\in$: belong(s) to

$Bn(n, \theta)$: binomial distribution with $n$ trials and probability $\theta$ of success in one trial

CDF: cumulative distribution function

CMC: conditional Monte Carlo

$\mathbb{C}or(X, Y) = \mathbb{C}ov(X, Y)/\sqrt{\mathbb{V}(X)\mathbb{V}(Y)}$: the correlation operator on $(X, Y)$

$\mathbb{C}ov(X, Y) = \mathbb{E}(X \cdot Y) - \mathbb{E}(X) \cdot \mathbb{E}(Y)$: the covariance operator on $(X, Y)$

CSP: constrained synchronized permutations

d.f.: degrees of freedom

EDF: empirical distribution function: $\hat{F}_{\mathbf{X}}(t) = \hat{F}(t|\mathcal{X}_{/\mathbf{X}}) = \sum_i \mathbb{I}(X_i \leq t)/n, t \in \mathcal{R}^1$

$\mathbb{E}(X) = \int_{\mathcal{X}} x \cdot dF_X(x)$: the expectation operator (mean value) of $X$

$\overset{d}{=}$: equality in distribution: $X \overset{d}{=} Y \leftrightarrow F_X(z) = F_Y(z), \forall z \in \mathcal{R}^1$

$\overset{d}{>}$: stochastic dominance: $X \overset{d}{>} Y \leftrightarrow F_X(z) \leq F_Y(z), \forall z$ and $\exists A : F_X(z) < F_Y(z), z \in A$, with $\Pr(A) > 0$

$<\neq>$: means $<$, or $\neq$, or $>$

$\sim$: distributed as, for example, $X \sim \mathcal{N}0, 1)$ means $X$ is standard normal distributed

$\forall$: for every

$F_X(z) = F(z) = \Pr\{X \leq z\}$: the CDF of $X$

$F_{X,Y}(z, t) = F(z, t) = \Pr\{X \leq z, Y \leq t\}$: the CDF of $(X, Y)$

i.i.d.: independent and identically distributed

$\mathbb{I}(A)$: the indicator function, that is, $\mathbb{I}(A) = 1$ if $A$ is true, and 0 otherwise

$\lambda = \Pr\{T \geq T^o|\mathcal{X}_{/\mathbf{X}}\}$: the attained $p$-value of test $T$ on dataset $\mathbf{X}$

$L_X(t) = L(t) = \Pr\{X \geq t\}$: the significance level function (same as the survival function)

$\mu = \mathbb{E}(\mathbf{X})$: the mean value of vector $\mathbf{X}$

MANOVA: multivariate analysis of variance

NOTATION AND ABBREVIATIONS

$\mathbb{M}d(X) = \tilde{\mu}$: the median operator on variable $X$ such that $\Pr\{X < \tilde{\mu}\} = \Pr\{X > \tilde{\mu}\}$

$n$: the (finite) sample size

$\mathcal{N}(\mu, \sigma^2)$: Gaussian or normal distribution with mean $\mu$ and variance $\sigma^2$

$\mathcal{P}(\tau)$: Poisson distribution with parameter $\tau$

$\Pr\{A\}$: a probability statement relative to $A \in \mathcal{A}$

$\mathcal{R}^n$: the set of $n$-dimensional real numbers

$\mathbb{R}$: the rank operator

$R_i = \mathbb{R}(X_i) = \sum_{1 \leq j \leq n} \mathbb{I}(X_j \leq X_i)$ the rank of $X_i$ within $\{X_1, \ldots, X_n\}$

$T^o = T_{obs} = T(X)$ the observed value of test statistic $T$ evaluated on $\mathbf{X}$

$\mathcal{U}(a, b)$: uniform distribution in the interval $(a, b)$

$\mathbb{V}(X) = \mathbb{E}(X - \mu)^2 = \sigma^2$: the variance operator on variable $X$

$X$ or $Z$: a univariate or multivariate random variable

$X$: a sample of $n$ units, $X = \{X_i, i = 1, \ldots, n\}$

$X^*$: a permutation of $X$

$|X| = \{|X_i|, i = 1, \ldots, n\}$: a vector of absolute values

$\biguplus$: the operator for pooling (concatenation) of two datasets: $X = X_1 \biguplus X_2$

# 1

# One- and two-sample location problems, tests for symmetry and tests on a single distribution

## 1.1 Introduction

Many real phenomena can be represented by numerical random variables. Considering a given population and a random sample of it, for forecasting or improving the effectiveness of inferential techniques related to estimation and testing of hypothesis, it would be useful to know the functional form of the distribution of the data. Sometimes, the central interest of the statistical analysis is focused only on the symmetry or on the location of the distribution itself. Another very common statistical problem consists of comparing two independent populations in terms of central tendency. In the simpler cases the object of the analysis is a univariate population, but in some real applications we are in the presence of many variables and multivariate datasets.

The methods presented in this chapter consist of rank or permutation procedures for the tests of the hypotheses cited above. Section 1.2 is an introduction to rank and permutation tests. In Section 1.3, devoted to one-sample tests, the Kolmogorov procedure for testing whether the data are distributed according to an hypothesized

*Nonparametric Hypothesis Testing: Rank and Permutation Methods with Applications in R,*
First Edition. Stefano Bonnini, Livio Corain, Marco Marozzi and Luigi Salmaso.
© 2014 John Wiley & Sons, Ltd. Published 2014 by John Wiley & Sons, Ltd.
Companion website: http://www.wiley.com/go/hypothesis_testing

cumulative distribution function (CDF), and the permutation test on the symmetry of the distribution are taken into account. Section 1.4 deals with multivariate one-sample tests, and introduces the multivariate location problem and the multivariate test on symmetry. In Section 1.5 the univariate two-sample location problem is discussed. Section 1.6 considers the multivariate extension of the location problem for two independent populations and presents some solutions for it.

In the one-sample problems the data are a random sample of numerical data $X = \{X_1, \ldots, X_n\}$ from the unknown population under study. In the two-sample problems the numerical sample data fom the $j$th unknown population are $X_j = \{X_{j1}, \ldots, X_{jn_j}\}$, with $j = 1, 2$ and $n_1 + n_2 = n$. In the multivariate extensions, in the presence of $q$ component variables, for the one-sample problem, the observation related to the $i$th statistical unit is denoted by $X_i = \{X_{i1}, \ldots, X_{iq}\}$ and, for the two-sample problem, the observation related to the $i$th statistical unit in the $j$th group is denoted by $X_{ji} = \{X_{ji1}, \ldots, X_{jiq}\}$.

## 1.2   Nonparametric tests

Traditional parametric testing methods are based on the assumption that data are generated by well-known distributions, characterized by one or more unknown population parameters (mean, median, variance, etc.) and the hypotheses of the problems are formulated as equalities/inequalities related to these unknown parameters. For example, the location problem can be formalized using the mean parameter, the scale problem can be expressed in terms of variance comparisons, etc.

In other words parametric methods are based on a modeling approach and on the introduction of stringent assumptions, often quite unrealistic, unclear and connected with the availability of inferential methods (Pesarin, 2001). Hence the critical values or alternatively the $p$-values can be computed according to the distribution of the test statistic under the null hypothesis, which can be derived from the assumptions related to the assumed underlying distribution of data. When the assumed distribution is not true, when we are not sure whether it is true or not or when it is not plausible, other methods, which ignore the true distribution of data, are needed. These methods are called *nonparametric* or *distribution-free*.

Since, when the parametric assumptions hold, the nonparametric procedures are only slightly less powerful than the parametric methods and they are the only valid solution when the parametric assumptions do not hold, nonparametric tests are in general more flexible and often more appropriate than parametric counterparts. Basically the nonparametric testing procedures can be classified into two kinds of methods: rank based tests and permutation tests.

### 1.2.1   Rank tests

The main aspect which characterizes rank tests is that observations are transformed into their sample ranks. Hence in the rank tests we compare the observations based on their ranks within the sample. Formally the rank of the $i$th observation with respect to a set of $n$ data is given by

$$R_i = \mathbb{R}(X_i) = \sum_{1 \leq j \leq n} \mathbb{I}(X_j \leq X_i),$$

where $\mathbb{R}$ is the (increasing) rank operator, $X_i$ is the transformed observation, $\mathbb{I}(A)$ is the indicator function of the event $A$, that is $\mathbb{I}(A) = 1$ when $A$ is true and $\mathbb{I}(A) = 0$ otherwise. Hence the rank of $X_i$ within $\{X_1, \ldots, X_n\}$ is equal to 1 if $X_i$ is the minimum value, it is equal to 2 if $X_i$ is the second smaller value, up to $n$ if $X_i$ is the maximum. Often in the case of ties, the midrank method is applied, that is the mean of the ranks corresponding to the positions in the sorted set of observations is assigned to the tied observations. Formally if rank $r$ is assigned to $t$ observations equal to a certain value $x$ ($t \leq r$), that is $r$ observations in the set $\{X_1, \ldots, X_n\}$ are less than or equal to $x$, then the rank of these $t$ observations, according to the midrank rule, is adjusted into the mean value of the $t$ ranks $(r - t + 1), (r - t + 2), \ldots, r$. Rank transformation is non-bijective, in the sense that a given set of ranks $\{R_1, \ldots, R_n\}$ may correspond to distinct sets of sample data.

Let us consider an example related to a pharmacological experiment. A pharmaceutical company needs to test whether a new experimental drug for lowering blood cholesterol levels is more effective than another drug already present in the pharmaceutical market. A group of patients is treated with the new drug and another group with the old drug. The null hypothesis consists of 'no difference' between the two treatment effects; the alternative hypothesis states the superiority of the new drug, that is the effect of the new drug is greater than the effect of the old one. Let us denote with $n_1$ and $n_2$ the number of patients treated with the new and the old drug, respectively, independent samples from populations with continuous probability function $F_1$ and $F_2$, respectively. The null hypothesis of no difference between the effects of the two treatments can be written as $H_0 : F_1 = F_2 = F$ with $F$ unknown. $H_0$ implies that the two samples can be considered as just one sample from a unique distribution $F$. A way to solve this problem is provided by the Wilcoxon rank sum test, a rank based testing procedure which takes into account the ranking of the observations within the pooled sample of $n_1 + n_2$ data and considers the sum of the ranks of the first sample as test statistic. When $H_0$ is true, the test statistic tends to take neither too large nor too small values. The distribution of the test statistic under the null hypothesis can be computed considering all the possible rankings as equally likely and the corresponding values of the statistic. Hence the computation of critical values and $p$-values does not depend on the unknown distribution $F$. This is why it is considered a distribution-free method.

## 1.2.2    Permutation tests and combination based tests

In many testing problems, the dataset can be seen as a partition into groups or samples according to the treatment levels of a real or symbolic experiment. According to the permutation testing principle, if two experiments characterized by the same sample space (the set of all possible samples) give the same dataset, then the result of the testing procedure conditional on the dataset itself must be the same, provided that the exchangeability condition with respect to samples holds under the null hypothesis (Pesarin, 2001). This is the reason why inference based on permutation tests is also called conditional inference.

In real applications, random sampling, on which the parametric methods are based, is rarely achieved. Hence often the unconditional inferences associated with parametric tests are not applicable. In these situations permutation tests are suitable solutions. Furthermore some common assumptions of parametric methods, such as the existence of mean values and variances, or equal variances of responses (homoscedasticity) under the alternative hypothesis are not needed within the permutation testing procedures.

For example, for the two-sample test related to the pharmaceutical problem, under the null hypothesis observations are exchangeable among samples because they are supposed to come from the same population and their belonging to one group or to another is actually random. A suitable test statistic for the problem may be the difference of the two-sample means which is expected to take neither too large nor too small values when $H_0$ is true. The distribution of the test statistic under the null hypothesis, and then the $p$-value of the test, can be computed considering all the possible permutations (i.e., reallocations of the observations to the two groups) as equally likely and computing the corresponding values of the statistic for each permutation. Alternatively, for computational simplicity, a random sample of all the possible permutations can be considered and the null distribution of the test statistic can be well approximated by Conditional Monte Carlo (CMC) techniques.

### 1.2.2.1 Nonparametric combination methodology

A suitable method to perform multivariate permutation tests or multiple permutation test procedures is the so called nonparametric combination (NPC) of dependent permutation tests. Let us suppose that the null hypothesis $H_0$ of a testing problem can be broken down into $k$ sub-hypotheses or partial hypotheses $H_{01}, \ldots, H_{0k}$ such that $H_0$ is true if and only if all the sub-hypotheses are true, formally $H_0 : \bigcap_{i=1}^{k} H_{0i}$. Similarly the alternative hypothesis $H_1$ is true if and only if at least one of the null sub-hypotheses is false, and consequently at least one of the alternative sub-hypotheses is true, briefly $H_1 : \bigcup_{i=1}^{k} H_{1i}$. Let $T = T(X)$ be a $k$-dimensional vector of test statistics and each component $T_i(X)$ be a suitable test statistic for testing $H_{0i}$ against $H_{1i}$ and without loss of generality assume that $H_{0i}$ is rejected for large values of $T_i(X)$. Assuming as usual that each row of the dataset corresponds to a statistical unit, and considering for example a test for independent samples, the NPC method works as follows:

1. Compute the vector of the observed values of $T$: $T_{obs} = [T_1(X), \ldots, T_k(X)]' = [T_{1(0)}, \ldots, T_{k(0)}]'$.

2. Consider a permutation of the rows of the dataset, that is a reallocation of the units to the groups, and compute the corresponding values of the test statistics: $\boldsymbol{T}^*_{(1)} = [T_1(\boldsymbol{X}^*_{(1)}), \ldots, T_k(\boldsymbol{X}^*_{(1)})]'$.

3. Perform $B$ independent repetitions of step (2) and obtain $\boldsymbol{T}^*_{(b)} = [T^*_{1(b)}, \ldots, T^*_{k(b)}]', b = 1, \ldots, B$.

4. For each $i$ compute an estimate of the significance level function $\Pr\{T^*_i \geq z\}$ : $\hat{L}_i(z) = \{\frac{1}{2} + \sum_r \mathbb{I}[T^*_{i(r)} \geq z]\}/(B+1), i = 1, \ldots, k$.

5. For each $b$ compute $\lambda^*_{i(b)} = \hat{L}_i(T^*_{i(b)}), b = 1, \ldots, B$ and compute $\lambda_{i(0)} = \hat{L}_i(T_{i(0)})$, $i = 1, \ldots, k$.

6. For each $b$ compute the combined values $T''^*_{(b)} = \psi(\lambda^*_{1(b)}, \ldots, \lambda^*_{k(b)})$ and $T''_{(0)} = \psi(\lambda_{1(0)}, \ldots, \lambda_{k(0)})$ using a suitable combining function $\psi$.

7. Compute the estimate of the $p$-value of the test as $\lambda'' = \sum_b \mathbb{I}[T''^*_{(b)} \geq T''_{(0)}]/B$.

The final decision should be based on $\lambda''$ in the sense that $H_0$ should be rejected in favor of $H_1$ if $\lambda'' \leq \alpha$. The NPC method is very useful to solve complex problems, in particular multivariate problems or problems where a multivariate test statistic may be suitable. The main advantage with respect to other standard parametric methods is that the multivariate distribution of the test statistic does not need to be known or estimated, and in particular the dependence structure between the component variables does not need to be known or explicitly specified. The dependence is implicitly taken into account through the permutation strategy and the application of the combining function $\psi$. The combining function must satisfy the following simple properties: (1) it must be nonincreasing in each argument; (2) it must attain its supremum even when only one argument tends to zero; and (3) for each $\alpha$ level the critical value $T''_\alpha$ is assumed to be finite and strictly smaller than the supremum value. Some suitable combining functions are:

- the Fisher *omnibus* combining funtion: $T''_F = -2 \sum_i \log(\lambda_i)$;

- the Liptak combining funtion: $T''_L = \sum_i \Phi^{-1}(1 - \lambda_i)$;

- the Tippett combining function: $T''_T = \max_i(1 - \lambda_i)$.

Tippett combination provides powerful tests when one or a few but not all of the alternative sub-hypotheses are true; Liptak's function has a more powerful behavior when all of the alternative sub-hypotheses are jointly true; Fisher's solution is intermediate between the two.

## 1.3  Univariate one-sample tests

The basic assumption of an inferential problem is that the observed phenomena can be represented by random variables with unknown distributions. The goal of the inferential study consists of investigating some aspects of the unknown distribution. Let us assume that the observed random sample has been drawn from a numerical population with unknown CDF $F(x)$. In order to test whether $F(x)$ is equal to a fully specified function (without any unknown nuisance parameter), a powerful and commonly used solution is provided by the procedure introduced by Kolmogorov (1933). Such a procedure is based on the comparison between the empirical distribution function (EDF) and the specified tested distribution (see Section 1.2.1). As it tests the distribution's fit to a set of data, it is classified as a *goodness-of-fit* test. In this sense it can be considered an alternative for ordinal data to the *goodness-of-fit* chi-square test, valid for nominal categorical variables. An important difference between the two procedures is that, for continuous variables, the Kolmogorov test is exact even for small sample sizes (in the case of non continuity it is not distribution-free), while the chi-square test requires that $n$ is large enough so that the test statistic under the null hypothesis approximately follows a chi-square distribution (Conover, 1999).

In some applications the test involves only one or a few aspects of the functional form of $F(x)$, hence only a specific property of $F(x)$ is specified under $H_0$. This is the case of the test on symmetry, very useful in particular in the statistical quality control of industrial processes (see Section 1.2.2). For continuous variables, symmetry of the distribution around the origin is equivalent to the property: $F(x) = 1 - F(-x)$ $\forall x \in \mathcal{R}$. Let us consider the cited one-sample problems.

### 1.3.1   The Kolmogorov goodness-of-fit test

Let $X = \{X_1, \dots, X_n\}$ be a random sample from a population with unknown continuous CDF $F(x)$ and assume an interest in testing the hypothesis that $F(x)$ corresponds to a known and completely specified distribution $F_0(x)$ against the alternative that this is not true. The testing procedure proposed by Kolmogorov (1933) is based on the supremum of the vertical distance between $F_0(x)$ and the EDF based on the observed sample $X$. Smirnov (1939) proposed an extension of the Kolmogorov test for comparing the distributions of two independent populations. Statistics based on the vertical distance between $F_0(x)$ and the EDF are called Kolmogorov-type statistics, while similar statistics based on the vertical distance between two EDFs are called Smirnov-type statistics (Conover, 1999). The Kolmogorov *goodness-of-fit* test presented in this paragraph is also called the one-sample Kolmogorov–Smirnov test. Formally the problem consists of testing the null hypothesis

$$H_0 : F(x) = F_0(x)$$

against the alternative

$$H_1 : F(x) \neq F_0(x).$$

The EDF of $X$ is $\hat{F}_n(x) = 1/n \sum_{i=1}^{n} \mathbb{I}(X_i \leq x)$ where $\mathbb{I}(X_i \leq x)$ takes value 1 if $X_i \leq x$ and 0 otherwise. The Kolmogorov test statistic is given by

$$D_n = \sup_{x \in \mathcal{R}} |\hat{F}_n(x) - F_0(x)|.$$

In some problems, the alternative hypothesis is one-sided, that is the CDF $F(x)$ is supposed to be smaller than $F_0(x)$ or larger than $F_0(x)$. Formally the one-sided alternative might be

$$H_1 : F(x) \leq F_0(x) \ \forall x \in \mathcal{R} \text{ and } F(x) < F_0(x) \text{ for some } x,$$

or similarly

$$H_1 : F(x) \geq F_0(x) \ \forall x \in \mathcal{R} \text{ and } F(x) > F_0(x) \text{ for some } x,$$

thus the suitable test statistic is $D_n^+ = \max_{x \in \mathcal{R}} [F_0(x) - \hat{F}_n(x)]$ and $D_n^- = \max_{x \in \mathcal{R}} [\hat{F}_n(x) - F_0(x)]$, respectively. The tests reject the null hypothesis for large values of the test statistics.

A result shows that if $X$ is a random sample from an absolutely continuous population with the CDF $F_0$, then the distribution of the statistic $D_n$ does not depend on $F_0$ but only on the sample size $n$ (Bagdonavicius et al. 2011). Therefore in the two-sided test, the hypothesis $H_0$ is rejected with a significance level $\alpha$ when $D_n > D_\alpha(n)$, where $D_\alpha(n)$ is the critical value of the statistic $D_n$, that is the $(1 - \alpha)$-quantile of the null distribution of $D_n$. Equivalently, the null hypothesis is rejected when the $p$-value of the test (probability that under $H_0$ the test statistic takes values greater than the observed value of $D_n$) is less than $\alpha$. A similar procedure should be applied to the one-sided tests. Exact quantiles for $D_n$ and approximate quantiles for $D_n^+$ and $D_n^-$ have been tabulated. When $n > 40$ the asymptotic approximation may be used. Some computationally friendly representations of the distribution of the test statistics for sample sizes less than 100 and with no ties are proposed by Marsaglia et al. (2003) and Birnbaum and Tingey (1951). When $F_0(x)$ is discrete, a modification for the computation of the quantiles of the test statistics might be applied (Conover, 1972; Coberly and Lewis, 1973).

The basic package of $R$ includes the function ks.test, which computes the Kolmogorov–Smirnov statistic for the one-sample or two-sample cases. The presence of ties in the case of noncontinuous variables generates a warning. If ties arise because of rounding, the test may be approximately valid, but even modest amounts of rounding can have an important effect on the computation of the test statistic.

Consider an industrial experiment in which we have a sample of $n = 10$ fabrics subjected to washing. The goal of the experiment is to analyze the performance of a new experimental detergent for clothes. Specifically, the response variable under study is the so called reflectance, that is the proportion of incident light which a given surface (of fabric) is able to reflect, which can be considered a measure related to the cleaning efficacy of the detergent. Suppose we wish to test, at the significance level

Table 1.1   Sample data of reflectance in the experiment on detergent performance.

| Piece of fabric | 1 | 2 | 3 | 4 | 5 |
|---|---|---|---|---|---|
| Reflectance | 0.608 | 0.533 | 0.912 | 0.498 | 0.885 |
| Piece of fabric | 6 | 7 | 8 | 9 | 10 |
| Reflectance | 0.291 | 0.805 | 0.436 | 0.868 | 0.721 |

$\alpha = 0.01$, whether the reflectance is uniformly distributed or not, namely whether it follows the distribution law $\mathcal{U}(0, 1)$. The observed sample is displayed in Table 1.1.

The null hypothesis of the testing problem is

$$H_0 : F(x) = F_0(x) = \begin{cases} 0 & \text{when} \quad x < 0 \\ x & \text{when} \quad 0 \le x < 1 \\ 1 & \text{otherwise,} \end{cases}$$

where $F(x)$ is the CDF of the reflectance, here represented by a continuous random variable. The function $F_0(x)$ specified in the null hypothesis is the CDF of the uniform distribution in the interval $[0, 1]$. The alternative hypothesis is $H_1 : F(x) \ne F_0(x)$.

The R code for the analysis is the following:

```
> ref=c(0.608,0.533,0.912,0.498,0.885,0.291,0.805,0.436,0.868,0.721)
> plot(ecdf(ref),xlim=c(0,1),verticals=TRUE,xlab="Reflectance",
    main="")
> curve(punif,from=0,to=1,add=TRUE,lty="dashed",lwd=2)
> ks.test(ref,"punif",alternative="two.sided")
```

and the output of ks.test is

```
####################################################################
#One-sample Kolmogorov--Smirnov test
#
#data: ref
#D = 0.336, p-value = 0.1651
#alternative hypothesis: two-sided
####################################################################
```

The command `plot(ecdf(ref),xlim=c(0,1),verticals=TRUE,xlab="Refle ctance",main="")` gives the EDF of the sample, and the argument `xlim` indicates the interval to be visualized on the $x$ axis. With the command `curve(punif,from=0,to=1,add=TRUE,lty="dashed", lwd=2)` we can draw on the same graph the CDF of the uniform distribution. The first argument indicates the type of probability distribution and `punif` corresponds to the $\mathcal{U}(0, 1)$ distribution.

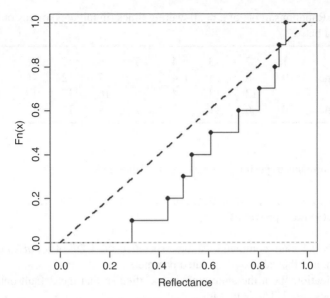

*Figure 1.1 Representation of the EDF of reflectance (dots) and CDF of $U(0, 1)$ (dashed line).*

The argument `lty` defines the line type. The graph is shown in Figure 1.1. It seems we are in the presence of an acceptable level of *goodness-of-fit*.

The first argument of the function `ks.test` indicates the observed data (`ref` in our example), the second one indicates the supposed distribution in the null hypothesis and the third indicates the type of alternative. The default is the two-sided alternative; the options `"less"` or `"greater"` correspond to the one-sided alternatives.

The observed value of the test statistic is $D = 0.336$ and it corresponds to the $p$-value 0.1651, thus there is no empirical evidence to reject the null hypothesis of uniform distribution for the reflectance.

When $F_0(x)$ is not continuous the Kolmogorov test tends to be conservative (Sprent and Smeeton, 2007) but, as we previously noted, methods for computation of $p$-values have been proposed also for discrete distributions. For example, let us consider the case of a bank where the average waiting time of a customer at the counter is equal to 8 min. We wish to test whether the waiting time follows a $\mathcal{P}(8)$ distribution, that is a *Poisson* distribution with parameter $\tau = 8$, by observing a random sample of $n = 20$ waiting times. The significance level is $\alpha = 0.05$. The null hypothesis is

$$H_0 : F(x) = F_0(x) = \begin{cases} 0 & \text{when} \quad x < 0 \\ \sum_{k=0}^{x} \dfrac{e^{-8}8^k}{k!} & \text{when} \quad 0 \le x < \infty \end{cases}$$

and the alternative is $H_1 : F(x) \ne F_0(x)$. The observed data are reported in Table 1.2.

Table 1.2    Random sample of $n = 20$ waiting times (in minutes) of customers at the counter in a bank.

| Customer | 1 | 2 | 3 | 4 | 5 | 6 | 7 | 8 | 9 | 10 |
|---|---|---|---|---|---|---|---|---|---|---|
| Waiting time | 9 | 8 | 5 | 4 | 4 | 7 | 12 | 6 | 9 | 11 |
| Customer | 11 | 12 | 13 | 14 | 15 | 16 | 17 | 18 | 19 | 20 |
| Waiting time | 10 | 6 | 11 | 7 | 7 | 8 | 11 | 13 | 9 | 9 |

The commands to perform the test are the following:

```
> wtime=c(9,8,5,4,4,7,12,6,9,11,10,6,11,7,7,8,11,13,9,9)
> ks.test(wtime,"ppois",8)
```

where in the `ks.test` command we have specified the CDF of a *Poisson* as tested distribution, and the value of the related parameter.

No indication about the alternative is specified so that the default option (two-sided) is considered. The output is

```
###############################################################
#One-sample Kolmogorov--Smirnov test
#
#data: x
#D = 0.2166, p-value = 0.305
#alternative hypothesis: two-sided
###############################################################
```

The observed value of the test statistic is 0.2166 and the $p$-value $= 0.305 > 0.05 = \alpha$ leads to the decision of no rejection of the null hypothesis that the distribution of the waiting time is $P(8)$. Note that in this situation a warning is generated, due to the presence of ties.

For denoting the specific distribution $F_0(x)$ in the null hypothesis of the Kolmogorov–Smirnov test, the options displayed in Table 1.3 can be used.

## 1.3.2    A univariate permutation test for symmetry

Sometimes an asymmetric distribution of the observed values of a response variable might be a symptom of abnormalities of the phenomena under study. For example, in the statistical quality control of industrial processes an asymmetry of the distribution of the response may reveal the presence of some problems in the manufacturing process. Let us assume that we are given a random sample of $n = 24$ washers drawn by the whole production of a metallurgical factory. The data consist of differences between the measured external diameters of the washers and the target value equal to 10 μm (Table 1.4).

Table 1.3  Options for the main distributions that can be tested with the Kolmogorov procedure.

| Discrete data | | Continuous data | |
|---|---|---|---|
| Option | Distribution | Option | Distribution |
| pbinom | Binomial | pbeta | Beta |
| pgeom | Geometric | pcauchy | Cauchy |
| phyper | Hypergeometric | pchisq | Chi-square |
| ppois | Poisson | pexp | Exponential |
| | | pf | F |
| | | pgamma | Gamma |
| | | pnorm | Normal |
| | | pt | Student's t |
| | | punif | Uniform |
| | | pweibull | Weibull |

Observing the density distribution of the observed data it is evident that we are in the presence of right or positive asymmetry (Figure 1.2) because some of the sampled washers present large external diameters (far from the target value) and this can cause a high percentage of waste. We are interested to test whether the whole production of the fabric is characterized by an asymmetric distribution of the external diameters at $\alpha = 0.10$.

By generalizing the problem, let $F$ be the unknown distribution of a continuous variable, and let $X = \{X_1, \ldots, X_n\}$ be a random sample of observations from such a population. The test on the symmetry of $F$ around the origin can be formalized by the hypotheses

$$H_0 : F(x) = 1 - F(-x)$$

against

$$H_1 : F(x) \neq 1 - F(-x).$$

To deal with this problem, we can consider the data as differences (Pesarin and Salmaso, 2010). In this case, the observed measurement of each unit is considered as if its sign was randomly assigned. In other words, under the null hypothesis the sign

Table 1.4  Differences (in micrometers) between the external diameters of washers and the target value (10 μm).

| | | | | | | | | | | |
|---|---|---|---|---|---|---|---|---|---|---|
| 1.6 | 1 | −0.8 | −1.3 | 1.4 | −0.1 | 1.1 | −1 | −0.1 | −0.6 | 0.7 | −0.6 |
| 2.1 | −1 | 3.5 | 0.6 | −0.2 | 0.5 | 0.5 | 4 | 1.9 | −0.4 | 0.4 | 1.4 |

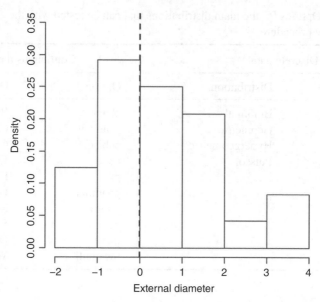

*Figure 1.2   Frequency histogram for the differences between the external diameters of washers and the target value (10 μm).*

of each difference is considered as if it were randomly assigned with probability $\frac{1}{2}$. Thus, one way to solve the testing problem is to consider the test statistic given by the absolute value of the sum of the sample values:

$$T = \left| \sum_{i=1}^{n} X_i \right|.$$

Its distribution $F_T(t|X)$, conditional to a given set of observed values $X_1, \ldots, X_n$, is obtained under the assumption that $H_0$ is true by considering the random attribution in all possible ways of the plus or minus sign to each value with equal probability. Operationally, considering $B$ permuted samples $\mathbf{X}^*$ obtained by randomly attributing the sign $+$ or $-$ to each $X_i$, and calculating for each permuted sample the corresponding permutation value $T^*$ of $T$, we can estimate the null distribution of $T^*$ according to the permutation distribution and compute the $p$-value of the test.

The $R$ code for the problem of the external diameters of washers follows:

```
> source("t2p.r")
> source("permsign.r")
> exdiam=c(1.6,1,-0.8,-1.3,1.4,-0.1,1.1,-1,-0.1,-0.6,0.7,-0.6,
+ 2.1,-1,3.5,0.6,-0.2,0.5,0.5,4,1.9,-0.4,0.4,1.4)
> exdiam=array(exdiam,dim=c(length(exdiam),1))
> perm.sign(x=exdiam,B=10000)
```

The output is

```
#######################################################################
# Permutation Test for Symmetry
# $p.value
# 0.04029597
#######################################################################
```

The source files for the execution of the test are "t2p.r" and "permsign.r" and they can be loaded with the source command. The former includes the code for the computation of the significance level function and the *p*-values; the latter contains the code for the computation of the test statistic and its permutation distribution. This is done with the function perm.sign which computes the statistic *T* and its permutation distribution, through random permutations of the signs of the observed data. The function can be used also for the two-sample test for paired data. In the case of testing for symmetry it requires two arguments: the array of observed data and the number *B* of random permutations of signs. The dataset (exdiam in this case) has to be set up as array with the array command. We consider $B = 10\,000$ permutations. The default value for the number of permutations for *B* is 1000. The function perm.sign invokes the function t2p that computes the permutation *p*-value as the frequency of permutation values of the test statistic that are greater than or equal to the observed one. It is worth noting that the *p*-value results from the permutation distribution of the statistic of interest that is obtained through random permutations of the signs of the data. Considering that we do not use all the possible permutations but, for computational convenience, only a random sample of *B* permutations, then performing the test many times, especially for small values of *B*, one can obtain different permutation distributions, hence different *p*-values. To remember the sequence of the considered sampled permutations the function set.seed(seed) can be used. With this function we can specify a seed to be associated with the set of sampled permutations and use this seed to identify and retrieve the same set of permutations when necessary. For example, if we repeat the analysis without specifying the seed we may obtain a different (but similar) *p*-value because, by performing the test again, a different sample of 10 000 permutations is drawn:

```
> perm.sign(exdiam, B=10000)
#######################################################################
# Permutation Test for Symmetry
# $p.value
# [1] 0.03779622
#######################################################################
```

For storing the set of sampled permutations and use it again, it is possible to specify the seed value (eg 1234) before the execution of the permutation test:

```
> set.seed(1234)
> perm.sign(exdiam, B=10000)
```

```
#####################################################################
# Permutation Test for Symmetry
# $p.value
# [1] 0.03969603
#####################################################################
```

From this moment, every time the command `perm.sign(exdiam,B=10000)` will be preceded by `set.seed(1234)`, the procedure will always return the $p$-value 0.03969603.

In this example, the $p$-value leads to the rejection of $H_0$ because it does not exceed the significance level $\alpha = 0.10$. The distribution of the external diameters of the whole production is not symmetric thus there is a problem in the production process. The distribution of the external diameters of the whole production is not symmetric and data agree with the hypothesis that there is a problem in the production process. For this reason let us operate a stratification of the data distinguishing the washers coming from plant A (first 12 values) and the ones produced by plant B (last 12 values). Figure 1.3 shows that the asymmetry of the distribution seems to be caused by the production of plant B, because it is not in target and produces a high percentage of washers with too large external diameter. This hypothesis must be tested with a two-sample test on location (see Section 2.4).

The test for symmetry may also be used for testing location on one-sample problems. To be specific, let us suppose that the observed data $Y_1, \ldots, Y_n$ are symmetrically distributed around $\delta$ and that $H_0 : \delta = \delta_0$, so that the transformations $X_i = Y_i - \delta_0$ are symmetrically distributed around 0 if and only if $H_0$ is true.

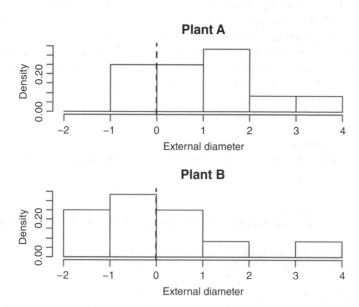

*Figure 1.3   Frequency histograms for the differences between the external diameters of washers and the target value (10 µm) by production plant.*

## 1.4    Multivariate one-sample tests

Consider the case of multivariate data. Let $X$ be a multivariate dataset from a sample of size $n$ and assume that the variable under study is $q$-dimensional. Formally the dataset is $X = \{X_{ih}; i = 1, \dots, n; h = 1, \dots, q\}$ where $X_{ih}$ denotes the $i$th observation of the $h$th variable. We assume that each of the $nq$-dimensional observations $\{X_{i1}, \dots, X_{iq}\}$ comes from a population with CDF $F_i(x; \theta)$, with $i = 1, \dots, n$, $x \in \mathcal{R}^q$ and $\theta = (\theta_1, \dots, \theta_q)'$ is a generic location (vector) parameter. In this section we consider one-sample tests concerning the location parameter $\theta$ and a multivariate extension of the test on symmetry presented in the previous section.

### 1.4.1    Multivariate rank test for central tendency

The random variable $Z$ taking values in $\mathcal{R}^q$ is said to be diagonally symmetric about $\mathbf{0}$ ($q$-dimensional vector of zeros) if both $Z$ and $-Z$ have the same CDF $F(z), z \in \mathcal{R}^q$. For absolutely continuous CDFs with density function $f(z)$ the diagonal symmetry can be represented by $f(z) = f(-z), \forall z \in \mathcal{R}^q$ (Puri and Sen, 1971).

Let us consider the multivariate location problem where the $n$ $q$-variate populations with CDFs $F_1(x; \theta), \dots, F_n(x; \theta)$ are independent and diagonally symmetric about $\mathbf{0}$. We wish to test whether the location (vector) parameter is null, that is $H_0 : \theta = \mathbf{0}$ against $H_1 : \theta \neq \mathbf{0}$. It is worth noting that the condition $F_1(x; \theta) = \dots = F_n(x; \theta)$ is not necessary (Puri and Sen, 1971).

For the general case of $q$-variate variables with not necessarily independent marginal components, let us consider the following transformation of $X$: $g(X) = \{s_i X_{ih}; i = 1, \dots, n; h = 1, \dots, q\}$ where $s_i = +1$ or $s_i = -1$. The number of possible vectors of signs $s = (s_1, \dots, s_n)'$ is $2^n$ hence, according to the basic permutation principle, the multivariate permutation distribution is spread over $2^n$ possible permutations. Under $H_0$ the permutation distribution is uniform because all the realizations are equally likely with probability equal to $2^{-n}$ (Puri and Sen, 1971). Hence we can obtain a distribution-free test for the present problem.

Let us now take into account a wide class of multivariate rank tests, useful to solve several different kinds of testing problems. Consider the $n \times q$ matrix $R = [R_{ih}]$ whose generic element $R_{ih}$ represents the rank of $|X_{ih}|$ in the set of values $\{|X_{1h}|, \dots, |X_{nh}|\}$. No ties are admissible because of the continuity assumption. For each variable (that is for each column of $R$) replace the ranks with the general scores $E^{(h)}(R_{ih})$. For each marginal variable consider the rank order statistics

$$T^{(h)} = \sum_{i=1}^{n} E^{(h)}\left(R_{ih}\right) c_{ih}, \quad \text{for } h = 1, \dots, q.$$

The weights $c_{ih}$ are the signs of the values $X_{ih}$, that is $c_{ih} = +1$ if $X_{ih} > 0$ and $c_{ih} = -1$ if $X_{ih} < 0$. Let us denote with $\mathbf{T}$ the $q$-dimensional vector of statistics $(T^{(1)}, \dots, T^{(q)})'$. According to the permutation distribution $E(\mathbf{T}) = \mathbf{0}$ and $Var(\mathbf{T}) = E(\mathbf{T}\mathbf{T}') = n\mathbf{V}$. The matrix $\mathbf{V} = [v_{jk}]$ is assumed to be positive definite (if singular it can be replaced by the highest order nonsingular minor of $\mathbf{V}$) with elements

$$v_{jk} = (1/n) \sum_{i=1}^{n} E^{(j)}\left(R_{ij}\right) E^{(k)}\left(R_{ik}\right) c_{ij} c_{ik}.$$

The test statistic for the multivariate location problem under study is

$$S = \frac{1}{n}\mathbf{T}'\mathbf{V}^{-1}\mathbf{T},$$

where $\mathbf{V}^{-1}$ is the inverse of $\mathbf{V}$. Large values of $S$ lead to the rejection of the null hypothesis in favor of the alternative hypothesis of non null central tendency.

According to the scores we can obtain different tests. Some examples are:

- $E^{(h)}(R) = 1, h = 1, \ldots, q$: multivariate sign test.

- $E^{(h)}(R) = R, \ \ h = 1, \ldots, q$: multivariate generalization of the one-sample Wilcoxon signed rank test.

- $E^{(h)}(R)$ is the expected value of the $R$th smallest observation of a sample of size $n$ from a chi-square distribution with 1 degree of freedom $h = 1, \ldots, q$: multivariate one-sample normal scores test.

We notice that all the considered tests cannot be applied to one-sided alternatives and they require the continuity assumption for the multivariate response variable. The solution proposed in this subsection for the present problem is the multivariate extension of the one-sample Wilcoxon signed rank test. The multivariate sign test does not require the symmetry, hence it is preferable to the signed rank test when this assumption is not realistic or not plausible. Furthermore, this is the only solution among these rank tests when only the signs of the differences are observed. Otherwise, under the symmetry assumption and when ranks of the sample differences can be determined, the Wilcoxon signed rank test is preferable because it uses more information than the sign test, and then it is more powerful under $H_1$, that is it rejects $H_0$ with higher probability when $H_0$ is not true. The normal score test is less flexible and it is preferable only in specific problems where it is reasonable to replace ordinary ranks with the related normal scores.

Consider a customer satisfaction survey about a recently opened shopping center. A sample of $n = 29$ customers was asked to evaluate 5 different aspects of the shopping center, such as the environmental temperature, the brightness, the presence of sales assistants, the range of products, and the background music volume. Note that these variables represent conditions that can make the shopping experience pleasant if present in the right amount, hence we can say that the best is 'neither too much nor too little'. Thus the evaluations are expressed on a scale from $-100$ ('too little') to $+100$ ('too much') and where 0 corresponds to 'just right'. We are interested to test if the mean values of the evaluations are equal to 0 or not at the significance level $\alpha = 0.05$. The sample data are reported in Table 1.5.

An $R$ function, to perform the multivariate one-sample location test based on ranks, is in the package ICSNP. To perform the analysis, the following commands should be typed:

```
> library(ICSNP)
> data=read.csv("mall.csv",header=TRUE,sep=";";)
> rank.ctest(X=data,Y=NULL,mu=NULL,scores="rank")
```

Table 1.5   Customer satisfaction survey of a new shopping center.

| Temperature | Brightness | Presence of sales assistants | Range of products | Background music volume |
|---|---|---|---|---|
| 20 | 61 | 35 | 57 | 58 |
| 42 | 40 | 17 | 11 | 9 |
| 38 | 22 | 46 | 36 | 12 |
| 0 | −1 | 10 | 16 | −11 |
| 100 | 19 | 30 | −31 | −25 |
| −20 | −41 | 47 | −14 | −94 |
| 5 | 18 | 43 | 14 | −26 |
| −5 | −21 | 34 | 48 | −45 |
| −40 | −78 | −68 | −10 | 13 |
| −61 | −82 | 25 | 7 | −88 |
| −83 | −76 | −71 | −89 | −72 |
| −77 | −84 | −30 | −58 | −86 |
| 99 | 59 | 56 | 44 | −92 |
| −79 | −75 | −36 | −24 | −73 |
| 3 | −3 | 27 | −22 | −12 |
| 41 | 4 | 50 | 26 | −9 |
| −37 | 23 | 29 | 45 | −46 |
| 21 | −18 | 8 | 33 | −34 |
| 60 | 62 | 64 | 65 | 66 |
| −19 | −42 | −44 | −13 | −23 |
| 2 | −6 | 15 | 32 | −93 |
| 98 | −29 | −17 | −63 | −43 |
| −2 | 6 | −38 | −33 | −15 |
| −80 | −85 | −87 | −56 | −70 |
| 1 | −4 | 63 | 31 | −27 |
| −60 | −74 | −59 | −16 | −39 |
| −81 | −62 | −90 | 28 | −57 |
| 39 | −7 | −8 | 37 | 24 |
| −35 | −65 | −91 | −69 | −28 |

The command library(ICSNP) is necessary to load the package ICSNP. Before installing ICSNP, the packages mvtnorm and ICS should be also loaded. The data of the present application can be loaded from the file mall.csv with the command data=read.csv("mall.csv",header=TRUE,sep=";"). The command rank.ctest(X=data,Y=NULL,mu=NULL,scores="rank") performs the test. The command requires a numeric data frame or matrix of data (X). The default value for the second argument Y is equal to NULL, thus a one-sample test is performed. The argument mu is a vector indicating the value of the location parameter under the null hypothesis. The default value is NULL, that represents the origin, thus for this problem we do not need to specify it. The argument scores indicates the score function we want to apply, and scores="rank" is the choice to perform a signed rank test. If scores="sign" a sign test is performed, whereas if scores="normal" a normal score test is performed.

The output is:

```
##########################################################################
# Marginal One Sample Signed Rank Test
#
# data: data
# T=15.7926, df=5, p-value=0.007462
# alternative hyp.: true location not equal to c(0,0,0,0,0)
##########################################################################
```

Thus the observed value of the test statistic is equal to 15.793 and the $p$-value of the test is equal to 0.007. Since the $p$-value is less than 0.05 we reject the null hypothesis in favor of the alternative that the vector of means is not equal to $(0, 0, 0, 0, 0)'$.

## 1.4.2    Multivariate permutation test for symmetry

Let us introduce now the multivariate extension of the permutation procedure for the test on symmetry. In this new problem the unknown distribution under investigation is multivariate. As stated in the univariate case, the problem of testing symmetry is a common problem in Statistical Quality Control where the goal of the analysis could be to test the symmetry of the distribution around the target of two or more characteristics of the product simultaneously considered. In other words, the interest is focused on the symmetry of the marginal distributions but without neglecting the multivariate nature of the problem and the possible dependence among the marginal variables. Formally let $f(x)$, with $x \in \mathcal{R}^q$, denote the joint probability function (for discrete variables) or density function (for continuous variables), and $f_i(x)$ the analogous marginal function of the $i$th component variable. The null hypothesis of the problem is $H_0 : f(x) = f(-x)$ and the alternative is $H_1 : f(x) \neq f(-x)$. Practically this is equivalent to test the diagonal symmetry of the multivariate distribution.

To face this problem we use the nonparametric combination (NPC) methodology (Pesarin and Salmaso, 2010). We consider the null hypothesis as the intersection of $q$ null sub-hypotheses of symmetry for each marginal distribution and we assume the global null hypothesis of symmetry to be true if each sub-hypothesis of marginal symmetry is true. Conversely the alternative hypothesis of the problem is true if at least one null sub-hypothesis is false. Hence the alternative hypothesis can be considered as the union of $q$ alternative sub-hypotheses of asymmetry. Formally, according to Roy's union-intersection principle (Roy, 1953), we can write the null hypothesis as

$$H_0 = \bigcap_{i=1}^{q} H_{0i}$$

Table 1.6   External and internal diameters of washers (difference from the target value in micrometers).

| External diameters | | | | | | | | | | | |
|---|---|---|---|---|---|---|---|---|---|---|---|
| 1.6 | 1 | −0.8 | −1.3 | 1.4 | −0.1 | 1.1 | −1 | −0.1 | −0.6 | 0.7 | −0.6 |
| 2.1 | −1 | 3.5 | 0.6 | −0.2 | 0.5 | 0.5 | 4 | 1.9 | −0.4 | 0.4 | 1.4 |
| Internal diameters | | | | | | | | | | | |
| −0.1 | 1 | 1.6 | 0.8 | 1 | −0.2 | 0.1 | 0.5 | −1.4 | −0.3 | 2.4 | −0.1 |
| 1.7 | 0.5 | −0.1 | 0.5 | 1.5 | 1.2 | 0 | 1.4 | 0.4 | −0.3 | 0.9 | −1.1 |

against the alternative

$$H_1 = \bigcup_{i=1}^{q} H_{1i},$$

where $H_{0i} : f_i(x) = f_i(-x)$ and $H_{1i} : f_i(x) \neq f_i(-x)$.

Under the null hypothesis exchangeability of the signs holds, that is for each $q$-dimensional vector of observations $(X_{i1}, \ldots, X_{iq})', i = 1, \ldots, n$, the signs can be permuted because the probability (for discrete variables) or density (for continuous variables) of observing $(X_{i1}, \ldots, X_{iq})'$ and $(-X_{i1}, \ldots, -X_{iq})'$ is the same. Let us consider again the industrial example of Section 1.3.2 and assume an interest in controlling both the external and internal diameters of washers. Let us assume to observe a random sample of $n = 24$ measures of differences from the target values of the external and internal diameters of washers drawn from the whole production. The data represent the difference of these measures (in micrometers) from the target values (Table 1.6 and Figure 1.4).

The $R$ commands to perform the test are:

```
> source("t2p.r")
> source("comb.r")
> source("permsign.r")
> exdiam=c(1.6,1,-0.8,-1.3,1.4,-0.1,1.1,-1,-0.1,-0.6,0.7,-0.6,
+ 2.1,-1,3.5,0.6,-0.2,0.5,0.5,4,1.9,-0.4,0.4,1.4)
> indiam=c(-0.1,1,1.6,0.8,1,-0.2,0.1,0.5,-1.4,-0.3,2.4,-0.1,
+ 1.7,0.5,-0.1,0.5,1.5,1.2,0,1.4,0.4,-0.3,0.9,-1.1)
> x=array(c(exdiam,indiam),dim=c(length(exdiam),2))
> perm.sign(x,fun="F",B=10000)
```

With the source() command use of the functions included in the files "t2p.r", "comb.r" and "permsign.r" is allowed. The code for performing the test is in "permsign.r"; the code for computing the significance level function is included in "t2p.r" and the code for the application of the NPC is in "comb.r". The perm.sign function for the multivariate case requires three arguments. The first argument is the dataset x ($n \times q$ matrix of data) defined as an array. The argument (fun) is the combination

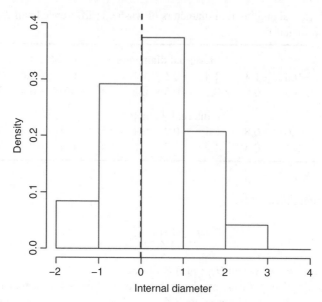

*Figure 1.4    Frequency histogram for the differences between the internal diameters of washers and the target value (in micrometers).*

function: "F", "L" and "T" represent Fisher, Liptak and Tippett's combination, respectively. The third argument represents the number of permutations for estimating the null permutation distribution of the test statistic.

The output, using the Fisher combining function and 10 000 permutations, is:

```
#####################################################################
# Multivariate Permutation Test for Symmetry
# Combination Function: Fisher
# $p.value
# 0.00549945
#####################################################################
```

Thus the *p*-value of the global test on symmetry is 0.005 and leads to reject the null hypothesis of symmetry of the multivariate distribution at the significance level $\alpha = 0.01$.

## 1.5    Univariate two-sample tests

In this section we address the problem of comparing two independent samples in the presence of one numerical variable. The data consist of $n = n_1 + n_2$ observations, where $n_j$ denotes the size of the *j*th sample ($j = 1, 2$). We are considering the most typical two-sample problem: the comparison of central tendencies.

Let us denote with $X_1 = \{X_{1i}, i = 1, \ldots, n_1\}$ the sample data from the first population and with $X_2 = \{X_{2i}, i = 1, \ldots, n_2\}$ the sample data from the second population. The most important nonparametric solutions for the two-sample location problem are the rank test proposed by Wilcoxon and the permutation two-sample test. These testing procedures are described in the following subsections.

## 1.5.1 The Wilcoxon (Mann–Whitney) test

The Wilcoxon rank sum test can be applied in the presence of independent random samples when the assumption of normal populations does not hold and the parametric $t$-test cannot be applied. The assumptions of this test are: (1) the data are realizations of continuous independent random variables and independence is assumed both between samples and within samples; (2) random variables generating data of the same sample are identically distributed. Moreover, the Wilcoxon test can also be applied in the presence of ordered categorical data, for example when the response variable represents categorical judges or it takes values in a Likert scale (Kvam and Vidakovic, 2007). The goal consists of comparing the central tendencies of the two samples, to test whether the locations of the respective populations are equal or not (two-sided test) or to test if one location is greater than the other (one-sided test). In some applications, especially (but not only) in medical or pharmacological studies, the problem consists of investigating the presence of a treatment effect represented by a shift of location. The null hypothesis is that of no treatment effect, that is, the samples can be thought as drawn from the same population.

An intuitive approach is to combine both samples into a single pooled ordered sample and then assign increasing ranks to the sample values, with no regard to which population each value comes from. Then the test statistic might be the sum of the ranks assigned to the first sample. Extreme values of the test statistic are empirical evidence in favor of the alternative hypothesis and the rejection region depends on the type of alternative, one-sided or two-sided (Conover, 1999).

By formalizing the problem, let us assume that the CDFs of the compared populations are $F_1(x)$ and $F_2(x)$ and let $X_{ji}$ be generated by the random variable $Z_{ji}$. There are several ways of specifying the one-sided test. Two of them are the stochastic (or random) effect model and the fixed effect model. According to the former we have $Z_{1i} = \mu + \Delta_{1i} + \epsilon_{1i}$ and $Z_{2i} = \mu + \epsilon_{2i}, i = 1, \ldots, n_j$, where $\mu$ is a constant, $\epsilon_{ji}$ $(j = 1, 2)$ are exchangeable random errors, with location equal to zero and scale parameter equal to $\sigma$, and $\Delta_{1i}$ are nonnegative random variables representing treatment effects. The latter is a special case where $\Delta_{1i} = \theta$ with probability one, with $\theta$ an unknown constant parameter. According to the fixed effects model the variances of the two compared populations are equal (homoscedasticity condition) and the two distributions may differ only in the location.

The one-sided test can be presented as a test on stochastic dominance or a test on location shift. In other words the hypothesis that, for example, the first population tends to assume greater values than the second one, can be represented as

$F_1(x) \leq F_2(x) \ \forall x \in \mathcal{R}$ and $F_1(x) < F_2(x)$ for some $x$ or, given the location parameter $\theta$, $F_1(x) = F_2(x - \theta)$ with $\theta > 0$. The null hypothesis of the problem is

$$H_0 : F_1(x) = F_2(x) \ \forall x \in \mathcal{R},$$

and the alternative hypothesis is the inequality between $F_1(x)$ and $F_2(x)$ just described or, according to the location-shift model we can write $H_0 : \theta = 0$ and $H_1 : \theta > 0$. The two representations are equivalent if we assume the fixed effect model. Henceforth in this subsection we will consider this model.

The test statistic of the Wilcoxon rank sum test (equivalent to the Mann–Whitney test) is based on the sum of the ranks of the elements of the first sample

$$W = \sum_{i=1}^{n_1} \mathbb{R}\left(X_{1i}\right),$$

where $\mathbb{R}(X_{1i})$ is the (increasing) rank of $X_{1i}$ in the pooled sample $\{X_{11}, \dots, X_{1n_1}, X_{21}, \dots, X_{2n_2}\}$, thus it is equal to 1 if $X_{1i}$ is the smallest observed value, to 2 if $X_{1i}$ is the second smallest observed value, and so on up to $n = n_1 + n_2$ for the largest value.

If the null hypothesis is true, the sum of the ranks of the first sample is expected to be similar to that of the second sample, hence when $W$ assumes large values $H_0$ should be rejected in favor of $H_1$. Under $H_0$ the distribution of the statistic $W$ does not depend on unknown parameters but depends on the sample sizes $n_1$ and $n_2$, because from the properties of ranks we obtain

$$P\left\{\left[\mathbb{R}\left(X_{11}\right), \dots, \mathbb{R}\left(X_{1n_1}\right)\right] = (j_1, \dots, j_{n_1})\right\} = \frac{n_2!}{n!}$$

for all $(j_1, \dots, j_{n_1})$ obtained from $n_1$ different elements of the set $(1, 2, \dots, n_1 + n_2)$. The minimum value of the statistic $W$ is $w_{\min} = 1 + \cdots + n_1 = n_1(n_1 + 1)/2$ and the maximum value is $w_{\max} = (n_2 + 1) + (n_2 + 2) + \cdots + (n_2 + n_1) = n_1(2n_2 + n_1 + 1)/2$. Hence for all $k = w_{\min}, \dots, w_{\max}$

$$P\{W = k\} = N_k \frac{n_2!}{n!}$$

where $N_k$ is the number of vectors $(j_1, \dots, j_{n_1})$ satisfying the condition $j_1 + \cdots + j_{n_1} = k$. The exact distribution of $W$ can be computed and tabulated. Clearly when the alternative hypothesis is that the the first population takes smaller values than the second, the null hypothesis is rejected for small values of the test statistic. Finally the two-sided alternative hypothesis of inequality in distribution should be rejected if $W \leq c_1$ or $W \geq c_2$, where $c_1$ and $c_2$ are the maximum natural number and the minimum natural number, respectively, verifying the inequalities $\sum_{k=w_{\min}}^{c_1} \Pr\{W = k | H_0\} \leq \alpha/2$ and $\sum_{k=c_2}^{w_{\max}} P\{W = k | H_0\} \leq \alpha/2$. Upper-tail probabilities of the Wilcoxon rank sum test statistic are available from www.wiley.com/go/hypothesis_testing (Hollander and Wolfe, 1999). Under $H_0$ the means and variances of the sum of ranks $W$ are

$$\mathbb{E}(W) = \frac{n_1(n+1)}{2}$$

and

$$\mathbb{V}(X) = Var(W) = \frac{n_1 n_2(n+1)}{12},$$

respectively. An important result (Bagdonavicius *et al.*, 2011) shows that if the probability distributions of the populations are absolutely continuous, then when $n \to \infty, n_1/n \to p \in (0,1)$ under the null hypothesis

$$Z_{(n_1,n_2)} = \frac{W - \mathbb{E}(W)}{\sqrt{\mathbb{V}(W)}} \xrightarrow{d} Z \sim N(0,1).$$

Hence for large sample sizes the normal approximation of the distribution of $W$ can be used. The Mann–Whitney test is a testing procedure proposed for the same location problems described. For the one-sided test, where the first population is supposed to take greater values than the second in the alternative hypothesis, the test statistic is

$$U = \sum_{i=1}^{n_1} \sum_{s=1}^{n_2} D_{is},$$

where $D_{is} = I\left(X_{1i} > X_{2s}\right) = 1$ if $X_{1i} > X_{2s}$ and 0 otherwise. With a similar logic the Mann–Whitney test for the lower-tail one-sided test and the two-sided test can be derived. The Mann–Whitney test is equivalent to the Wilcoxon rank sum test.

The basic package of $R$ contains the function `wilcox.test`, which computes the Wilcoxon rank sum test statistic and the $p$-value for the one-sample and the two-sample case. By default an exact $p$-value is computed if the sample sizes are less than 50 and there are no ties. Otherwise, a normal approximation is used.

Let us consider the following problem. Before being able to enrol in a first level degree course of Economics at some Italian Universities, students have to do an entrance test related to mathematical skills. The examination consists of a written test and, according to the final score, the students could be asked to participate to a preliminary remedial course. In Table 1.7 the test results for a university of two samples of candidates coming from scientific and classical studies backgrounds, respectively, are shown. Sample sizes are $n_1 = n_2 = 10$. We wish to test whether the mathematical skills of the two groups of students are equivalent against the alternative hypothesis that students coming from scientific studies are better prepared in Mathematics. Let $Score_{scient}$ and $Score_{class}$ denote the random variables representing the test result for a student from a scientific and from a classical high school, respectively. The hypotheses of the testing problem are $H_0 : Score_{scient} \overset{d}{=} Score_{class}$ and $H_1 : Score_{scient} \overset{d}{>} Score_{class}$.

Table 1.7   Results of the examination of mathematical skills for applicants enrolling in a university Economics course, coming from scientific and classical studies backgrounds.

| Scientific studies | | | | | | | | | |
|---|---|---|---|---|---|---|---|---|---|
| 82.261 | 81.191 | 74.902 | 87.119 | 84.410 | 81.551 | 90.806 | 82.818 | 71.843 | 82.504 |

| Classical studies | | | | | | | | | |
|---|---|---|---|---|---|---|---|---|---|
| 66.131 | 89.327 | 75.119 | 68.449 | 77.942 | 70.756 | 68.533 | 65.219 | 82.723 | 66.637 |

The symbol $\overset{d}{=}$ denotes equality in distribution, that is equality of the CDFs. The mathematical notation $\overset{d}{>}$ denotes stochastic dominance, that is the cumulative distribution of $Score_{scient}$ is less than or equal to the CDF of $Score_{class}$ (and the strict inequality is true for some subsets of $\mathcal{R}$). In other words, under the alternative hypothesis the scores of candidates coming from scientific studies tend to be distributed on greater values. The significance level is $\alpha = 0.01$.

Looking at the two-sample density histograms (Figure 1.5), it seems that the score distribution of students from scientific studies is shifted toward greater values than the score of students from classical studies. For testing whether this conclusion based on descriptive statistics can be extended to the corresponding populations we may apply the Wilcoxon–Mann–Whitney test.

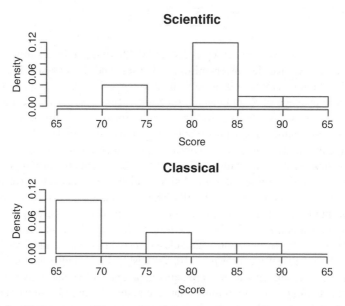

*Figure 1.5   Histograms of the results of the examination of mathematical skills for applicants enrolling in a university Economics course, coming from scientific and classical studies backgrounds.*

The R code for the analysis is the following:

```
> scient=c(82.261,81.191,74.902,87.119,84.410,81.551,90.806,
+ 82.818,71.843,82.504)
> class=c(66.131,89.327,75.119,68.449,77.942,70.756,68.533,
+ 65.219,82.723,66.637)
> wilcox.test(scient,class,alternative="greater")
```

Through the command wilcox.test(x,y,alternative="greater") we can per-form the one-sided test of Wilcoxon–Mann–Whitney for comparing $x$ and $y$ (vectors of data of the first and second sample, respectively) for testing the hypothesis that the first population tends to take greater values than the second. For testing the opposite one-sided hypothesis the last argument should be alternative="less". For the two-sided alternative of inequality in distribution the syntax is alternative= "two.sided".

The final output after the application of the R code to the described problem is:

```
######################################################################
#Wilcoxon rank sum test
#
#data: scient and class
#W = 81, p-value = 0.009272
#alternative hypothesis: true location shift greater than 0
######################################################################
```

The observed value of the test statistic is $W = 81$ and the corresponding $p$-value is equal to 0.009 that leads to reject the null hypothesis in favor of the hypothesis that the scores of candidates coming from scientific studies tend to be greater at the significance level $\alpha = 0.01$.

A problem, similar from the statistical point of view but related to a completely different application, is the following. An experiment is designed to see if farmed fish exhibit a lower protein content than wild fish caught in the open sea. The experiment is performed on a species of saltwater fish. The goal consists of assessing whether there is a significant negative difference between the percentages of proteins in farmed fish and in wild fish. Let $Prot_{farmed}$ denote the percentage of proteins in farmed fish and $Prot_{sea}$ denote the percentage of proteins in wild fish. The null hypotheses of the problem is $H_0 : Prot_{farmed} \overset{d}{=} Prot_{sea}$ and the alternative is $H_1 : Prot_{farmed} \overset{d}{<} Prot_{sea}$. Two samples of healthy fish, similar in terms of age, gender, weight, etc., of sizes $n_1 = n_2 = 12$ were drawn from the respective populations (Table 1.8).

From a descriptive point of view sample data related to farmed fish tend to be greater than data related to the other sample, as shown in Figure 1.6.

Then the R code for this test is:

```
> farm=c(18.85,16.93,19.29,18.31,17.27,18.64,17.82,19.00,19.58,
+ 18.04,17.27,19.19)
```

Table 1.8   Percentage of proteins in two samples of farmed and wild fish.

| | | | | | Farmed fish | | | | | | |
|---|---|---|---|---|---|---|---|---|---|---|---|
| 18.85 | 16.93 | 19.29 | 18.31 | 17.27 | 18.64 | 17.82 | 19.00 | 19.58 | 18.04 | 17.27 | 19.19 |
| | | | | | Wild fish | | | | | | |
| 19.23 | 19.57 | 19.50 | 18.64 | 18.70 | 19.54 | 19.04 | 20.67 | 20.71 | 18.99 | 19.37 | 19.06 |

```
> sea=c(19.23,19.57,19.50,18.64,18.70,19.54,19.04,20.67,20.71,
+ 18.99,19.37,19.06)
> wilcox.test(farm,sea,alternative="less")
```

and the output is:

```
###########################################################################
#Wilcoxon rank sum test
#
#data: farm and sea
#W = 26.5, p-value = 0.004672
#alternative hypothesis: true location shift less than 0
###########################################################################
```

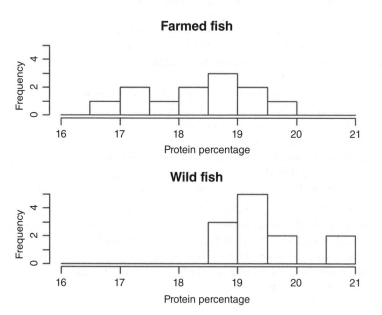

*Figure 1.6   Histograms of protein percentage in two samples of farmed and wild fish.*

The option for the alternative hypothesis now is `"less"`. The value of the test statistic is $W = 26.5$ and it corresponds to a $p$-value equal to 0.005, that leads to rejecting the null hypothesis in favor of the hypothesis of stochastic dominance, that is the amount of proteins in farmed fish is lower.

### 1.5.2 Permutation test on central tendency

The problem of comparing the central tendency of two independent samples in the presence of one numerical variable may be addressed also through a permutation solution. Let us assume homoscedasticity (i.e., equal variances) in the null hypothesis and denote with $F_1$ and $F_2$ the compared nondegenerate continuous distributions, both from the same family $F$. Consider the stochastic dominance problem where in the alternative hypothesis the first population is supposed to take greater values than the second. In other words, $H_1$ asserts the stochastic dominance of the first population on the second. Note also that $H_0$ implies exchangeability of observed data with respect to groups, and observed data may be viewed as if they were randomly assigned to two groups but they come from the same population.

The permutation solution does not need the assumption that means and variances of the response variables are finite. It only needs location parameters (mean, median, or others) to be finite and proper sampling indexes for them to be available. Unlike the Wilcoxon test, the permutation test does not require the continuity of the response variables and can be applied also in the presence of ties without any correction or approximation. Furthermore rank transformation is not one-to-one with respect to the dataset $X$, hence the sufficiency property is not satisfied. A transformation of $X$ is a sufficient statistic if it contains all the necessary information for solving the inferential problem on $F$. Hence the Wilcoxon rank sum test can have some power decay. Instead the permutation test is conditioned to the whole dataset $X$ which is a sufficient statistic for $F$ (Pesarin, 2001).

Let us consider $X = X_1 \uplus X_2$ with $\uplus$ denoting vector concatenation, so that the two samples are pooled into one and the first $n_1$ elements of $X$ correspond to $X_1$ and the remaining $n_2$ elements to $X_2$.

A suitable permutation test statistic is $T^* = \overline{X}_1^* - \overline{X}_2^*$ where $\overline{X}_j^* = \sum_{i=1}^{n_j} X_{ji}^*/n_j$, $j = 1, 2$, are the sample means of the first $n_1$ elements and of the remaining $n_2$ elements of $X^*$, respectively, and $X^*$ is a permuted dataset, that is a vector obtained by changing the position of elements, or equivalently by randomly assigning $n_1$ of the observed values to the first sample and the remaining to the second. As a consequence of exchangeability, under the null hypothesis the distribution of $T^*$ can be estimated by permuting the dataset $B$ independent times and computing the value of the statistic corresponding to each permutation (Pesarin and Salmaso, 2010). The $p$-value of the test is $\lambda = \sum_{b=1}^{B} \mathbb{I}(T_{(b)}^* \geq T^0)/B$ (proportion of $T^*$ permuted values greater than or equal to $T^0$), where $T_{(b)}^*$ is the value of the statistic related to the $b$th permutation and $T^0$ is the observed value of the statistic corresponding to the unpermuted dataset.

Consider the problem of the examination of mathematical skills of applicants enrolling in a university Economics course, coming from scientific and classical studies backgrounds (Table 1.7 and Figure 1.5). The *R* code for the application of the permutation test and the final result are:

```
> source("t2p.r")
> source("perm_2samples.r")
> data=c(scient,class)
> lab=rep(1:2,each=10)
> data_mat=cbind(lab,data)
> T=perm.2samples(data_mat,alt="greater",B=10000)
> T$p.value
[1] 0.00779922
```

The function perm.2samples(data,alt,B), defined in the file "perm_2samples.r" (which requires the code in the file "t2p.r" for the computation of the *p*-value), computes the permutation distribution of the test statistic and it requires that data are arranged in a matrix where the first column contains the label of the groups and the second column contains the observed data. We can specify the type of alternative as usual with the options alt="greater", alt="less" or alt="two.sided". The default number of permutations is $B = 1000$ but a different number can be specified in the third argument (e.g., $B = 10\,000$). Here the matrix of data is data_mat:

|        | lab | data   |
|--------|-----|--------|
| [1,]   | 1   | 82.261 |
| [2,]   | 1   | 81.191 |
| ...    | ... | ...    |
| [10,]  | 1   | 82.504 |
| [11,]  | 2   | 66.131 |
| [12,]  | 2   | 89.327 |
| ...    | ... | ...    |
| [20,]  | 2   | 66.637 |

The *p*-value is computed with the function t2p that computes the significance level function of the test statistic *T* according to the permutation distribution. The vector t2p(T) contains the significance levels corresponding to each of the permutation values of the test statistic and the first element is the significance level corresponding to the observed value of the test statistic, that is the *p*-value. The *p*-value can be obtained by typing T$p.value. The value 0.008 leads to the rejection of the null hypothesis in favor of the alternative that candidates from scientific schools are better prepared in mathematics. The result is then the same of the Wilcoxon test.

Also for the second problem of fish (Section 2.4.1), we can apply the permutation test:

```
> data=c(farm,sea)
> lab=rep(1:2,each=12)
> data_mat=cbind(lab,data)
```

```
> T=perm.2samples(data_mat,alt="less",B=10000)
> T$p.value
[1] 0.00089991
```

Even in this case the $p$-value (0.001) is less than $\alpha$, hence the null hypothesis must be rejected in favor of the alternative hypothesis that the amount of proteins in farmed fish is less than in fish coming from the sea. The conclusion is then similar to that of the Wilcoxon test.

## 1.6    Multivariate two-sample tests

This section is dedicated to the multivariate extension of the two-sample location problem. The dataset here consists of $n = n_1 + n_2$ observations from two independent $q$-variate populations, where $n_1$ and $n_2$ are the sizes of the two samples. The $i$th multidimensional observation in the $j$th group is denoted by $X_{ji} = \{X_{ji1}, \dots, X_{jiq}\}$. A rank based and a permutation solution are considered.

### 1.6.1    Multivariate tests based on rank

Let us consider the problem of testing the identity of two multivariate distributions $F_1$ and $F_2$. We shall assume that $F_1$ and $F_2$ have a common unspecified form but possible different location vectors.

Let $F_j(x)$ be the CDF, belonging to the class of all continuous distribution functions, for the $j$th population with $j = 1, 2$. According to the fixed effect model and denoting with $Z_{ji}$ the multivariate random variable from which $X_{ji}$ is assumed to be generated, we have $Z_{1i} = \mu + \theta + \epsilon_{1i}$ and $Z_{2i} = \mu + \epsilon_{2i}$, $i = 1, \dots, n_j, j = 1, 2$, where $\mu \in \mathcal{R}^q$ is a constant vector, $\epsilon_{ji}$ are exchangeable $q$-variate random errors, with location equal to the null vector and variances/covariances matrix equal to $\Sigma$, and $\theta \in \mathcal{R}^q$ is the vector of parameters representing treatment fixed effects. The hypothesis to be tested is

$$H_0 : F_1(x) = F_2(x) = F(x) \text{ for all } x \in \mathcal{R}^q$$

against the general two-sided alternative

$$H_1 : F_1(x) \neq F_2(x).$$

For the translation-type alternatives, the null hypothesis can be written as

$$H_0 : \theta = 0$$

against the alternative

$$H_1 : \theta \neq 0.$$

The alternative hypothesis states that at least one element of the vector of effects $\theta$ is not equal to zero, that is at least for one component of the response variable we have a non-null effect.

Let $R_{jih}$ be the rank of $X_{jih}$ in the set $\{X_{11h}, \ldots, X_{1n_1 h}, X_{21h}, \ldots, X_{2n_2 h}\}$ for $h = 1, \ldots, q$. Let $R_h$ denote the observed $n$-dimensional vector of ranks related to the $h$th variable and consider the $n \times q$ matrix $R = [R_1, \ldots, R_q]$. Each column of this matrix can be considered as a permutation of the numbers $1, 2, \ldots, n$. Thus $R$ can be considered a realization of a $n \times q$ random matrix with $(n!)^q$ possible realizations. Since the $q$ marginal components of the multivariate response are in general stochastically dependent, the joint distribution of the elements of the matrix of ranks will depend on the unknown distribution $F$ even when $H_0 : F_1(x) = F_2(x) = F(x)$ holds. Under $H_0$, the distribution of the matrix of ranks conditional to the observed matrix $R$ over the set $S(R^*)$ of all the possible permutations of the rows $R^*$ under $H_0$ is uniform. The probability of observing one of the $n! = n \cdot (n-1) \cdot \ldots \cdot 2 \cdot 1$ possible realizations is $1/n!$.

A general class of rank scores can be defined as follows:

$$E^{(h)}(R) = g_h \left( \frac{R}{n+1} \right),$$

with $h = 1, \ldots, q$ and $1 \le R \le n$. Hence the matrix $R$ can be replaced by $E = [E_1, \ldots, E_q]$ where

$$E_h = \left[ E^{(h)}\left(R_{11h}\right), \ldots, E^{(h)}\left(R_{1n_1 h}\right), E^{(h)}\left(R_{21h}\right), \ldots, E^{(h)}\left(R_{2n_2 h}\right) \right]'$$

with $h = 1, \ldots, q$. For each sample and for each of the $q$ variables the average rank score can be computed as

$$T_{j \bullet h} = \frac{\sum_{i=1}^{n_j} E^{(h)}\left(R_{jih}\right)}{n_j}.$$

Under $H_0$ the average rank scores should be close to the total mean scores $\overline{T}_{\bullet \bullet h} = \left( n_1 \overline{T}_{1 \bullet h} + n_2 \overline{T}_{2 \bullet h} \right)/n$ and the contrasts $\left( \overline{T}_{j \bullet h} - \overline{T}_{\bullet \bullet h} \right)$ should stochastically be close to zero for $j = 1, 2$ and $h = 1, \ldots, q$. In the presence of $C \ge 2$ groups a suitable test statistic for this problem might be

$$L = \sum_{j=1}^{C} n_j [\overline{T}_j - \overline{T}]' V^{-1} [\overline{T}_j - \overline{T}],$$

where $\overline{T}_j = [\overline{T}_{j \bullet 1}, \ldots, \overline{T}_{j \bullet q}]'$, $\overline{T} = [\overline{T}_{\bullet \bullet 1}, \ldots, \overline{T}_{\bullet \bullet q}]'$ and $V$ is the permutation covariance matrix of the contrasts $\overline{T}_j - \overline{T}$ under $H_0$ (see Section 3.5). $L$ is Hotelling's type test statistic used in a parametric test which assumes that data are generated by a multivariate normal distribution. Hence it is a suitable test statistic under normality.

When the multivariate distribution is very far from the normal, this statistic may be not a valid choice. Hence for the two-sample problem a possible test statistic is

$$L = n_1[\overline{T}_1 - \overline{T}]'V^{-1}[\overline{T}_1 - \overline{T}] + n_2[\overline{T}_2 - \overline{T}]'V^{-1}[\overline{T}_2 - \overline{T}].$$

For large values of $n$ and $q$ asymptotic distributions for $L$ are derived (Puri and Sen, 1971). The permutation distribution of $L$ asymptotically, in probability, reduces to the chi-square distribution with $q(C - 1)$ degrees of freedom. Thus for $C = 2$ the null hypothesis should be rejected when $L \geq \chi^2_{q;\alpha}$ where $\alpha$ denotes the significance level. According to the type of scores, different tests can be performed:

- For the *Multivariate Multisample Median Test* the score should be

$$E^{(h)}(R) = \begin{cases} 1 & \text{if } R \leq n/2 \\ 0 & \text{otherwise,} \end{cases}$$

  hence $T_{j\bullet h}$ is the proportion of values less than the median in the $j$th sample for the $h$th variable.

- For the *Multivariate Rank Sum Test* we define

$$E^{(h)}(R) = \frac{R}{n+1},$$

  hence the statistic $T_{j\bullet h}$ is equal to $(n + 1)^{-1}$ times the average rank in the $j$th sample for the $h$th variable.

- For the *Normal Scores Test* we put $p = \frac{n-R+1}{n+1}$ and

$$E^{(h)}(R) = z_p,$$

  where $z_p$ is the $(1 - p)$th quantile of the standard normal distribution that is the value such that $\Phi(z_p) = 1 - p$, with $\Phi$ denoting the CDF of the standard normal distribution.

The median test is preferable when the interest is focused on median comparisons, that is when the median is the location parameter under study. The normal scores test may be used only in specific problems, when it is reasonable to replace ordinary ranks with the related normal scores. Otherwise, among the rank based procedures, the rank sum test is the better choice. All the procedures described in the present subsection are based on the assumption that responses are continuous variables and can be applied only for two-sided tests.

Let us consider again the example related to the entrance test to enrol in the Economics course (Subsection 2.4.1). In this application, the scores of a sample of 20 candidates to enroll in the Economics course are related to mathematical skills and to economic knowledge. Half of the 20 students come from scientific studies and the others come from classical studies. We want to test whether the bivariate distribution of the scores of the two groups are the same, against the alternative, that the distributions of the two groups differ. In other words the goal consists of testing

Table 1.9    Results of the examination of mathematical skills and economic knowledge for applicants enrolling in a university Economics course, coming from scientific and classical studies backgrounds.

| | Scientific studies | | | | |
|---|---|---|---|---|---|
| Mathematical skills | 82.261 | 81.191 | 74.902 | 87.119 | 84.41 |
| Economic knowledge | 87.807 | 96.851 | 77.155 | 99.330 | 98.570 |
| Mathematical skills | 81.551 | 90.806 | 82.818 | 71.843 | 82.504 |
| Economic knowledge | 69.909 | 75.220 | 62.405 | 73.750 | 81.182 |
| | Classical studies | | | | |
| Mathematical skills | 66.131 | 89.327 | 75.119 | 68.449 | 77.942 |
| Economic knowledge | 79.451 | 92.708 | 74.730 | 66.063 | 62.818 |
| Mathematical skills | 70.756 | 68.533 | 65.219 | 82.723 | 66.637 |
| Economic knowledge | 92.883 | 99.869 | 97.991 | 61.801 | 84.395 |

whether the mathematical skills and the economic knowledge of the groups are the same or not. The significance level is $\alpha = 0.10$. Formally by denoting with *Math* and *Econ* the variables representing the scores of the two tests, the null hypothesis is

$$H_0 : (Math_{scient}, Econ_{scient}) \overset{d}{=} (Math_{class}, Econ_{class}),$$

that is the null hypothesis is true if the bivariate distributions of the scores on Mathematics and Economics are equal between the groups. The alternative hypothesis is

$$H_1 : (Math_{scient}, Econ_{scient}) \overset{d}{\neq} (Math_{class}, Econ_{class}),$$

that is the scores of the two groups are different. Table 1.9 shows the data for the problem.

The *R* package ICSNP contains tools for nonparametric multivariate analysis. In particular in this package is the function rank.ctest(X,Y,mu,scores) that performs the $C$-sample location test (with $C \geq 2$) based on marginal ranks, for which the three described score functions are available. For the two-sample test the function requires the $n_j \times q$ matrices of sample observations (X,Y) and a vector indicating the difference in the means under the null hypothesis (mu). NULL indicates no difference between the group means. The argument scores requires the type of score test to be performed to be specified. It may be "sign" for a sign test, "rank" for a rank test or "normal" for a normal score test. The *R* code for the analysis is:

```
> library("ICSNP")
> data=read.csv("test_eco.csv",header=TRUE,sep=";")
> X=data[1:10,c(2,3)]
> Y=data[11:20,c(2,3)]
> rank.ctest(X,Y,mu=NULL,scores="rank")
```

The output is:

```
###################################################################
#    Marginal Two Sample Rank Sum Test
#
#data: X and Y
#T=5.8684, df=2, p-value=0.05317
#alternative hypothesis: true location difference is
#    not equal to c(0,0)
###################################################################
```

The function returns some information and in particular the observed value of the test statistic $T$ and the $p$-value for the test (p-value). Note that only the two-sided alternative is available for this test. Thus for our multivariate example, we observe a $p$-value equal to $0.053 < 0.10$ that leads to the rejection of the null hypothesis of equal distributions in favor of the alternative that the two score distributions are different.

Another interesting problem is the multivariate extension of the application on fish also introduced in Section 2.4. Let us assume an interest in assessing whether there is a difference between farmed and wild fish in terms of percentage of proteins and lean body mass (lbm). The hypothesis we want to test is that there is no difference in the bivariate distribution between the two groups, against the general alternative that the two distributions differ. The observed data on the two samples of fish are reported in Table 1.10. The hypotheses of the problem are.

$$H_0 : \left(Prot_{farmed}, LBM_{farmed}\right) \overset{d}{=} \left(Prot_{sea}, LBM_{sea}\right),$$

against

$$H_1 : \left(Prot_{farmed}, LBM_{farmed}\right) \overset{d}{\neq} \left(Prot_{sea}, LBM_{sea}\right).$$

Table 1.10   Percentages of proteins and lean body mass in two samples of farmed and wild fish.

| | Wild fish | | | | | |
|---|---|---|---|---|---|---|
| Proteins | 20.67 | 19.34 | 18.67 | 19.33 | 19.42 | 19.80 |
| Lean body mass | 5.62 | 3.32 | 4.10 | 8.50 | 5.94 | 4.45 |
| Proteins | 19.01 | 18.91 | 18.51 | 19.09 | 18.99 | 19.63 |
| Lean body mass | 8.24 | 7.90 | 5.60 | 1.97 | 10.50 | 5.50 |
| | Farmed fish | | | | | |
| Proteins | 17.27 | 18.55 | 19.03 | 18.03 | 17.17 | 18.63 |
| Lean body mass | 0.98 | 2.74 | 7.82 | 1.33 | 1.56 | 5.47 |
| Proteins | 17.82 | 18.40 | 19.22 | 19.32 | 19.11 | 19.08 |
| Lean body mass | 7.98 | 5.20 | 1.78 | 1.23 | 4.78 | 5.91 |

Under the null hypothesis, the percentage of proteins and lbm levels are equal in the two groups of fish while under the alternative the two groups of fish differ.

The $R$ code to perform the analysis is:

```
> library("ICSNP")
> bass_mv=read.csv("bass_mv.csv",header=TRUE,sep=";")
> X=bass_mv[1:12,c(2:3)] #first sample
> Y=bass_mv[13:24,c(2:3)] #second sample
> rank.ctest(X,Y,scores="rank")
```

The output is:

```
########################################################################
#    Marginal Two Sample Rank Sum Test
#
#data: X and Y
#T=8.8108, df=2, p-value=0.01221
#alternative hypothesis: true location difference is
#          not equal to c(0,0)
########################################################################
```

The resulting $p$-value is 0.012 hence at the significance level $\alpha = 0.05$ the null hypothesis should be rejected in favor of the alternative that the two populations of fish differ in the amount of proteins and/or lbm.

## 1.6.2    Multivariate permutation test on central tendency

A natural extension of the two-sample permutation test to multivariate problems is now presented. In this framework the data is $q$-dimensional ($q \geq 2$). Often in tests for complex hypotheses, in the presence of many response variables or when several aspects of the distribution are involved, the overall testing problem can be broken down into a finite set of $k > 1$ different partial tests. Note that the number $q$ of responses does not always coincide with $k$, although for most multivariate location problems $k = q$. As in the problems considered before, the null hypothesis consists of the equality in distribution of two multivariate responses, for example the equality in distribution of each marginal variable. The NPC methodology can be applied. Even in this case $H_0$ may be properly and equivalently broken down into a finite set of sub-hypotheses $H_{0i}$, $i = 1, \ldots, k$ each appropriate for a partial aspect of interest or for a marginal variable (Pesarin and Salmaso, 2010). Therefore $H_0$, also called the global null hypothesis, is true if all the $H_{0i}$ are jointly true and thus it may be written as

$$H_0 : \bigcap_{i=1}^{k} H_{0i}.$$

The alternative hypothesis states that at least one of the null sub-hypotheses $H_{0i}$ is not true. Hence the alternative may be represented by the union of $k$ sub-alternatives as

$$H_1 : \bigcup_{i=1}^{k} H_{1i}$$

where each sub-hypothesis $H_{1i}$ is the alternative of $H_{0i}$. Thus $H_1$ is true when at least one sub-alternative is true. In this framework, $H_1$ is called the global alternative hypothesis. For each univariate partial test on central tendency the difference of sample means may be a suitable test statistic and a univariate permutation test on central tendency may be applied (Section 1.5.2). Through the NPC methodology (Section 1.2.2) the partial $p$-values are combined to obtain a univariate test statistic suitable to solve the multivariate problem.

The main advantage of this procedure, besides the possibility of considering multivariate response variables neither assuming any specific distribution nor specifying the dependence structure among the component variables (but taking it into account implicitly), is in its great flexibility that allows the solution of very complex problems. No continuity assumption is needed, hence it may be applied to continuous, discrete or mixed multivariate variables. It can also be applied to one-sided alternatives and even to complex alternatives where some of the partial sub-hypotheses $H_{1i}$ are two-sided and others one-sided with different possible directions.

Let us consider the example of the examination for students wishing to enroll in the Economics course. The scores in the Mathematics examination and in the Economics examination are considered for the sample of 20 students, 10 of which come from scientific studies and the others come from classical studies. To test whether the distributions of the two populations of students are the same, against the alternative that the distribution of the population coming from a scientific high school is stochastically greater, that is that students from scientific studies tend to get better results, using the multivariate permutation test with $B = 5000$ permutations, the following $R$ code should be applied:

```
> source("dataperm.r")
> source("umultiaspect.r")
> source("t2p.r")
> source("comb.r")
> data=read.csv("test_eco.csv",header=TRUE,sep=";")
> lab=rep(c(1,2),c(10,10))
> data_rev=cbind(lab,data[,2:3])
> perm=dataperm(dataset=data_rev,B=5000)
> l=u_multi_aspect(perm,rep("DM",2),rep(1,2),maspt=0)$P
> T2=comb(l,"F")
> pv=t2p(T2)[1]
> pv
> [1] 0.01898102
```

The scripts are included in the files "dataperm.r", "umultiaspect.r", "t2p.r" and "comb.r" which can be loaded with the source command. The first is useful for permutating the dataset, the second for the calculation of the permutation multivariate distribution of the partial test statistics, the third for obtaining the significance level function and the fourth for the combination of partial tests. After the loading of data from the file "test_eco.csv" the dataset must be set in the form:

```
    var1   var2
1   x11    x12
1   x21    x22
... ...
1   xn1    xn2
2   y11    y12
2   y21    y22
... ...
2   ym1    ym2
```

that is with the vector of labels of the two groups (of sizes $n$ and $m$) in the first column and with the variables of interest in the following columns.

The function dataperm(dataset,B) performs $B$ random permutations of the multivariate dataset obtaining $B$ permuted datasets. The function u_multi_aspect considers the observed and permuted datasets (output of dataperm), computes the multivariate distribution of partial test statistics and the corresponding significance levels according to the type of alternative. In this case the test statistics are the differences of means ("DM") for both the partial tests (i.e., for both the variables) hence the second argument is rep("DM",2). In the case of categorical variables, instead of difference of means, it is possible to use the Anderson–Darling statistic with the option "AD". In the present problem the type of alternative, for each partial test, is *group1* > *group2* hence the next argument is rep(1,2) or equivalently c(1,1). If we want consider the opposite alternative *group2* > *group1* for both the partial tests we have to specify rep(-1,2) or c(-1,-1) and for the two-sided alternative rep(0,2) or c(0,0). Of course different alternatives can be specified and tested for the two partial tests using c(-1,1), c(1,0), etc. To recover the multivariate significance level function the command is l=u_multi_aspect(...)$P. Thus we can obtain the combined test statistic combining the significance level functions and compute the corresponding $p$-value through the commands T2=comb(l,"F") and l2=t2p(T2)[1], respectively. For the combination, the possible choices are "F" for the Fisher function, "L" for the Liptak rule and "T" for the Tippett formula.

The resulting $p$-value is $0.019 < \alpha = 0.10$ hence, as with the rank test, the null hypothesis should be rejected in favor of the alternative of better preparation of students coming from scientific high schools. The $p$-value of the permutation test is less than that of the rank test (0.053). This result is consistent with the greater power of the combination based procedure.

# References

Bagdonavicius, V., Kruopis, J., Nikulin, M.S. (2011) Non-parametric Tests for Complete Data. John Wiley & Sons, Ltd.

Birnbaum, Z.W., Tingery, F.H. (1951) One-sided confidence contours for probability distribution functions. The Annals of Mathematical Statistics, 22, 4, 592–596.

Coberly, W.A., Lewis, T.O. (1973) A note on the one-sided Kolmogorov–Smirnov test of fit for discrete distribution functions. Annals of the Institute of Statistical Mathematics, 24, 183–187.

Conover, W.J. (1972) A Kolmogorov goodness-of-fit test for discontinuous distributions. Journal of the American Statistical Association, 67, 591–596.

Conover, W.J. (1999) Practical Nonparametric Statistics. John Wiley & Sons, Ltd.

Hollander, M., Wolfe, D.A. (1999) Nonparametric Statistical Methods. John Wiley & Sons, Ltd.

Kolmogorov, A.N. (1933) Sulla determinazione empirica di una legge di distribuzione. Giornale dell'Istituto Italiano degli Attuari, 4, 83–91.

Kvam, P.H., Vidakovic, B. (2007) Nonparametric Statistics with Applications to Science and Engineering. John Wiley & Sons, Ltd.

Marsaglia, G., Wai Wan, T., Jingbo, W. (2003), Evaluating Kolmogorov's distribution. Journal of Statistical Software, 8, 18.

Pesarin, F. (2001). Multivariate Permutation Tests with Applications in Biostatistics. John Wiley & Sons, Ltd.

Pesarin, F., Salmaso, L. (2010) Permutation Tests for Complex Data: Theory, Applications and Software. John Wiley & Sons, Ltd.

Puri, M.L., Sen, P.K. (1971) Nonparametric Methods in Multivariate Analysis. John Wiley & Sons, Ltd.

Roy, S.N. (1953) On a heuristic method of test construction and its use in multivariate analysis. Annals of Mathematical Statistics, 24, 220–238.

Smirnov, N.V. (1939) On the estimation of the discrepancy between empirical curves of distribution for two independent samples. Bulletin Moscow University, 2, 3–16.

Sprent, P., Smeeton, N.C. (2007) Applied NonParametric Statistical Methods. Chapman & Hall/CRC.

# 2

# Comparing variability and distributions

## 2.1 Introduction

In this chapter, the data consist of two independent random samples, one sample
drawn from each of two underlying populations. We would like to make inference
about the possible presence of difference other than pure location ones between two
populations. Section 2.1 introduces the problems to be addressed and the method
presented in the successive sections.

Section 2.2 deals with nonparametric methods for detecting the difference in
variability. Problems concerning comparison of variability are of interest in quality
control, in agricultural production systems, in experimental education and in many
other areas. Tests for the equality of variances are often of interest also as a preliminary
to analysis of variance, dose–response modeling or discriminant analysis. Comparing
scales is much harder than comparing locations. There are two reasons for this and
these will be discussed in Section 2.2. In Section 2.2.1 we present the Ansari–
Bradley rank test which is the most familiar rank test for comparing variability. This
test requires that the medians of the populations underlying the samples are equal.
In Sections 2.2.2 and 2.2.3 we present two permutation tests which do not require
this assumption: the permutation Pan test, which is a version of the Levene test, is
particularly useful when the distributions of the populations underlying the samples
of interest are heavy-tailed and/or highly skewed; and the permutation O'Brien test
which is particularly useful for symmetric and light-tailed distributions. Practical
applications in industry quality control and in experimental ecology are discussed.

Section 2.3 deals with nonparametric methods for jointly detecting difference
in location and scale. It is important to note that in many biomedical situations the

*Nonparametric Hypothesis Testing: Rank and Permutation Methods with Applications in R*,
First Edition. Stefano Bonnini, Livio Corain, Marco Marozzi and Luigi Salmaso.
© 2014 John Wiley & Sons, Ltd. Published 2014 by John Wiley & Sons, Ltd.
Companion website: http://www.wiley.com/go/hypothesis_testing

treatment can change location and scale simultaneously. The two-sample location-scale problem arises also in other fields such as climate dynamics and stock prices. Since in practice, normal data as well as either pure location or pure scale alternatives are not common, we address the location-scale problem within the permutation/rank framework where normal data are not required. The most familiar rank test is due to Lepage and is presented in Section 2.3.1. In Section 2.3.2 we consider also the Cucconi test, proposed earlier but not nearly as well known. The methods are applied to a biomedical case-control study.

Finally, Section 2.4 deals with nonparametric methods for detecting the general difference between the distribution functions of the underlying populations. Here the two-sample problem is addressed from the most general point of view because we would like to assess whether there are any differences between the parent population distributions. The tests presented, respectively, in Sections 2.4.1 and 2.4.2 are the Kolmogorov–Smirnov test and the Cramer–von Mises test that are sensitive to all possible types of differences between the two parent population distributions, including differences in shape, and thus in kurtosis and/or skewness. Therefore, these tests should be used when it is suspected that differences in shape as well as differences in location and/or scale (between the two distributions) are possible. A practical economic application is discussed.

Data: $n = n_1 + n_2$ observations: $X_1 = \{X_{1i}, i = 1, \ldots, n_1\}$ and $X_2 = \{X_{2i}, i = 1, \ldots, n_2\}$.

Assumptions

A1: $X_1$ and $X_2$ are random samples from a continuous population 1 and 2, respectively.

A2: $X_1$ and $X_2$ are mutually independent.


## 2.2   Comparing variability

Problems concerning comparison of variability are of interest in many areas. Minimizing and controlling variability is important in quality control (Nair and Pregibon, 1988), in agricultural production systems (Fairfull et al., 1985), in experimental education (Games et al., 1972; Olejnik and Algina, 1988). Tests for the equality of variances are often of interest also as a preliminary to analysis of variance, dose–response modeling or discriminant analysis, but a strong argument should be made in these cases. For example, the SAS TTEST procedure computes the $F$ test for the equality of variances as a preliminary for choosing between the pooled variance $t$ test and the Satterthwaite or Cochran test (SAS Online documentation 9.2). It is well-known that the common $F$ test is extremely non robust to non-normality, so it does not make much sense to use a non robust variance procedure before a relatively robust procedure like the $t$ test. Box (1953) emphasized that 'to make the preliminary test on variances is rather like putting to sea in a rowing boat to find out whether conditions are sufficiently calm for a ocean liner to leave port!'. The $F$ test is not robust to departures from normality even asymptotically (Box, 1953). Unlike the Student $t$ test, the $F$ test is very sensitive to non-normality: for short-tailed distributions, the test is conservative, while more seriously, for long-tailed distributions the type-one error rate is much larger than the nominal level. Moreover, its level of significance is very sensitive also to departures with respect to symmetric distributions (Tan, 1982). It is important to emphasize that comparing variances or other measures of scale is much harder than comparing means or other measures of location. There are two reasons for this (Boos and Brownie, 2004). The first reason is that the normal theory test statistics for detecting location differences are standardized to be robust to non-normality, and so the test procedures are approximately exact. This is not true for the normal theory test statistics for detecting scale differences, which are not asymptotically distribution free, but depend on the kurtosis of the parent distributions. The second reason is that for location comparisons the hypothesis that the populations may differ only in location is often appropriate allowing the use of permutation methods that are exact; on the contrary, for scale comparisons the hypothesis that the populations may differ only in scale rarely makes sense since usually location differences are allowed. Given that, it is necessary to adjust for unknown means or locations by subtracting means or other location measures, but the transformed data are not exchangeable and then permutation tests are no longer exact (Good, 2000).

There are many comparative simulation studies of tests for scale differences in the literature. See, among others, Hall (1972), Brown and Forsythe (1974), Geng *et al.* (1979), Keselman *et al.* (1979), Conover *et al.* (1981), Balakrishnan and Ma (1990) and Lim and Loh (1996). A recent study by Marozzi (2011) proposed for the first time in the literature the resampling versions of several Levene type tests. Two of them are presented in Sections 2.2.2 and 2.2.3.

## 2.2.1    The Ansari–Bradley test

A measurable characteristic of a raw material must have some specified average value, but the variability should also be small to keep the characteristics of the end product within specifications, and so it is central to determine whether two samples of products have significantly different variability or not. The weight (kg) variability of two samples of lead ingots from two different distributors has been analyzed. Table 2.1 displays the dataset (Gibbons and Chakraborti, 2003). Ingots are used as raw material in an industrial production process.

Each distributor has a specification of median weight of 16 kg (note that sample medians and means of the two samples are just 16 kg), but it is suspected that the first

Table 2.1    Weight (kg) of lead ingots from two distributors.

| First distributor | | | | | | | | | | |
|---|---|---|---|---|---|---|---|---|---|---|
| 15.4 | 16 | 15.6 | 15.7 | 16.6 | 16.3 | 16.4 | 16.8 | 15.2 | 16.9 | 15.1 |

| Second distributor | | | | | | | | | | |
|---|---|---|---|---|---|---|---|---|---|---|
| 15.7 | 16.1 | 15.9 | 16.2 | 15.9 | 16 | 15.8 | 16.1 | 16.3 | 16.5 | 15.5 |

distributor distributes ingots with a greater weight variability. To test this hypothesis we may use the Ansari–Bradley test, which is a rank procedure for comparing two independent samples as far as variability is concerned under the assumption

A3: the medians of the populations underlying the samples are equal.

Let $F_1$ and $F_2$ be the distribution functions corresponding to populations 1 and 2, respectively. The null hypothesis is that population 1 and population 2 have the same unknown distribution

$$H_0 : \{F_1(x) = F_2(x), \forall x \in \mathcal{R}\}.$$

Let us consider the location-scale model which corresponds to taking

$$F_1(x) = H\left(\frac{x - \mu_1}{\sigma_1}\right) \text{ and } F_2(x) = H\left(\frac{x - \mu_2}{\sigma_2}\right), \forall x \in \mathcal{R},$$

where $H$ is the distribution function of a continuous variable with median 0. Therefore, $\mu_1$ and $\mu_2$ are the medians (location parameters) of populations 1 and 2, respectively; $\sigma_1$ and $\sigma_2$ are the scale parameters of populations 1 and 2, respectively.

The system of hypotheses under testing can be restated as

$$H_0 : \{\sigma_1 = \sigma_2\} \text{ against } H_1 : \{\sigma_1 \neq \sigma_2\}$$

when the alternative hypothesis is two-sided. In the ingots of raw material example, the alternative hypothesis is one-sided $H_1 : \{\sigma_1 > \sigma_2\}$ because it assumes that the first distributor distributes ingots with greater weight variability.

The Ansari–Bradley test is based on the following statistic

$$AB = \sum_{i=1}^{n_1} AB_i,$$

where $AB_i$ is the $AB$ score assigned $X_{1i}$. To compute the $AB$ scores, first sort the combined sample elements in increasing order; second assign the score 1 to both the smallest and largest elements, the score 2 to the second smallest and second largest elements, etc. The $AB$ scores are

$$1, 2, 3, ..., \frac{n}{2}, \frac{n}{2}, ..., 3, 2, 1 \text{ when } n \text{ is even, and}$$

$$1, 2, 3, ..., \frac{n-1}{2}, \frac{n+1}{2}, \frac{n-1}{2}, ..., 3, 2, 1 \text{ when } n \text{ is odd.}$$

The $AB$ statistic is the sum of the scores assigned to $X_1$ elements. Note that the Ansari–Bradley test can be applied also when populations 1 and 2 have different medians, provided that they are known, by subtracting $\mu_1$ to the first sample elements and $\mu_2$ to the second sample elements because the median adjusted variables satisfy

Assumption A3. The motivation of the test is that if for example $\sigma_1 > \sigma_2$, then $\mathbf{X}_1$ elements tend to be more spread out that $\mathbf{X}_2$ elements. Therefore $AB$ scores of $\mathbf{X}_1$ elements would be tendentially smaller than $AB$ scores of $X_1$ elements and so the $AB$ statistic. The contrary happens if $\sigma_1 < \sigma_2$. Therefore the one-sided alternative hypothesis $H_1 : \{\sigma_1 > \sigma_2\}$ is rejected for small values of the $AB$ statistic; the one-sided alternative hypothesis $H_1 : \{\sigma_1 < \sigma_2\}$ is rejected for large values of the $AB$ statistic; the two-sided alternative hypothesis $H_1 : \{\sigma_1 \neq \sigma_2\}$ is rejected either for small or large values of the $AB$ statistic. Prior to the widespread of powerful and relatively cheap computers, to apply the Ansari–Bradley test one had to look at the corresponding table of critical values or to consider its large sample approximation (Hollander and Wolfe, 1999). Moreover, in the presence of ties, the formula for computing it is rather tedious. Nowadays, using software like $R$, the test can be applied very easily, without concern about looking at critical values or ties or large sample approximations.

The basic package of $R$ contains the function `ansari.test` which computes both the $AB$ statistic and the $p$-value. If both samples contain less than 50 finite values and there are no ties the exact $p$-value is computed. Otherwise the asymptotic normal approximation is used. Note that in the presence of ties mid-ranks are used. The $R$ code for analyzing the lead ingots data follows.

```
> #first distributor
> x1=c(15.4,16,15.6,15.7,16.6,16.3,16.4,16.8,15.2,16.9,15.1)
> #second distributor
> x2=c(15.7,16.1,15.9,16.2,15.9,16,15.8,16.1,16.3,16.5,15.5)
> ansari.test(x1,x2,alternative="greater")
###################################################################
#         Ansari--Bradley test
#
#data:   x1 and x2
#AB = 46.5, p-value = 0.007005
#alternative hypothesis: true ratio of scales is greater than 1
###################################################################
```

Note that $R$ warns about the presence of ties and then computes the $p$-value using the asymptotic approximation. Since the $p$-value of the Ansari–Bradley test is 0.007 there is strong practical evidence to reject the hypothesis of equal variability in favor of that of greater variability for the first distributor. We may conclude that the first distributor distributed lead ingots with greater weight variability with respect to the second distributor. Therefore the end product producer should prefer the first distributor in order to keep the characteristics of the end product within specifications more easily. It is important to note that if you would like to compare the result using $R$ and the result using the table for critical values as in Hollander and Wolfe (1999), you have to take care that $R$ computes the $AB$ statistic on the first sample rather than on the second sample as done in Hollander and Wolfe (1999) and then the test speaks against the null hypothesis for large (rather than small) values of the test statistic.

To obtain the *AB* statistic write

```
> ansari.test(x1,x2,alternative="greater")$statistic
AB 46.5
```

and to obtain the p-value of the *AB* test write

```
> ansari.test(x1,x2,alternative="greater")$p.value
[1] 0.00700511
```

In the case of the opposite one-sided alternative $H_1 : \{\sigma_1 < \sigma_2\}$ just replace `alternative="greater"` with `alternative="less"`. In the case of the two-sided alternative $H_1 : \{\sigma_1 \neq \sigma_2\}$, which is the default in *R*, use `alternative="two.sided"`. It is worth noting that the additional package COIN contains a more general function named `ansari_test`. We have left the reader to go into this general function thoroughly since for the purpose of this subsection the basic function `ansari.test` suffices.

Note that the Ansari–Bradley test can be used for analyzing the lead ingots example because each distributor has a specification of median weight of 16 kg, and in fact the medians of the two samples are just 16 kg. When Assumption A3 is not satisfied the use of the Ansari–Bradley test is strongly discouraged (unless you apply the test to the median adjusted samples) because it may display a bizarre behavior as in the case when the maximum of $X_1$ elements is less than the minimum of $X_2$ elements, that is no overlapping between the samples. If for example $n_1 = 6$ and $n_2 = 5$, for all possible $X_1$ and $X_2$, the *AB* statistic is always equal to 21, regardless of the values of $\sigma_1$ and $\sigma_2$. This means that in such a situation the Ansari–Bradley test is completely ineffective to test for possible difference in the scale parameters. When Assumption A3 is not met, we suggest to use the methods presented in the next two subsections. It is important to note that Assumption A3 is not strictly necessary for the methods presented in Sections 2.2.2 and 2.2.3 which are more general than the Ansari–Bradley test. The reader may wonder why the Ansari–Bradley test is even considered in this book if there are better alternatives. There are two main reasons. The first reason is that the Ansari–Bradley test is still the most familiar rank test for comparing variability. The second reason is that it is one of the two components of the most familiar nonparametric test for jointly detecting location and scale differences (see Section 2.3.1).

## 2.2.2  The permutation Pan test

The data displayed in Table 2.2 give the times in minutes to breakdown of an insulating fluid under elevated voltage stresses of 32 and 36 kV, respectively, for the first and second sample. Fifteen observations are taken at each voltage (Nair, 1984). The experimenter would like to answer this question: is the variability of times significantly higher for the 32 kV power voltage? It might be suspected that parent distributions are highly skewed because observations are failure times. This is a situation in which the *F* test is very likely not to be appropriate.

Table 2.2   Time (in minutes) to breakdown of an insulating fluid under voltages of 32 and 36 kV.

| 32 kV | | | | | | | |
|---|---|---|---|---|---|---|---|
| 0.27 | 0.40 | 0.69 | 0.79 | 2.75 | 3.91 | 9.88 | 13.95 |
| 15.93 | 27.80 | 53.24 | 82.85 | 89.29 | 100.58 | 215.50 | |

| 36 kV | | | | | | | |
|---|---|---|---|---|---|---|---|
| 0.35 | 0.59 | 0.96 | 0.99 | 1.69 | 1.97 | 2.07 | 2.58 |
| 2.71 | 2.90 | 3.67 | 3.99 | 5.35 | 13.77 | 25.50 | |

The Ansari–Bradley test is also not appropriate because Assumption A3 is not tenable: both means and medians are very different. The R code to obtain means and medians is:

```
> #32k volt sample
> x1=c(.27,.40,.69,.79,2.75,3.91,9.88,13.95,15.93,27.80,53.24,82.85,
89.29,100.58,215.50)
> #36k volt sample
> x2=c(.35,.59,.96,.99,1.69,1.97,2.07,2.58,2.71,2.90,3.67,3.99,5.35,
13.77,25.50)
> mean(x1)
[1] 41.18867
> median(x1)
[1] 13.95
> mean(x2)
[1] 4.606
> median(x2)
[1] 2.58
```

A test for the comparison of variability which is appropriate in this case is the permutation version of the Pan (1999) test proposed by Marozzi (2011). The Pan test is a modification of the Levene test which is the most familiar alternative to the $F$ test. The original Levene test itself has several versions. To introduce the Pan test we consider the version of the original Levene test found to be the best by Brown and Forsythe (1974) and Conover et al. (1981). It is simply the Student $t$ test on $Z_{ji} = |X_{ji} - \tilde{X}_j|$ where $\tilde{X}_j$ is the median of the $j$th sample. Contrary to the original Levene test, this one is asymptotically distribution-free (Miller, 1968), robust (Brown and Forsythe, 1974) and powerful when the underlying distributions have long tails (O'Brien, 1979). If $n_1$ is odd, the structural null absolute deviation from the sample median is discarded and the test statistic should be modified by considering $n_1 - 1$ in place of $n_1$. The same happens if $n_2$ is odd. This version of the Levene test possesses robustness, efficiency, simplicity and nice asymptotics, but Pan (1999) found a flaw: in some situations its power never reaches 1, no matter how different $\sigma_1$ and $\sigma_2$ are.

This leads Pan (1999) to propose a modification to remove this problem but retain the nice features of the test. The Pan test is based on the following statistic

$$PAN = \frac{\ln \overline{Z}_1 - \ln \overline{Z}_2}{\sqrt{\frac{1}{n_1}\frac{V_1^2}{\overline{Z}_1^2} + \frac{1}{n_2}\frac{V_2^2}{\overline{Z}_2^2}}},$$

where $\overline{Z}_j$ and $V_j^2$ denote, respectively, the mean and the variance of $Z_{ji}, i = 1, ..., n_j, j = 1, 2$. $H_0 : \{\sigma_1 = \sigma_2\}$ is rejected at the $\alpha$ level of significance in favor of the two-sided $H_1 : \{\sigma_1 \neq \sigma_2\}$ if $|PAN| \geq t_{1-\alpha/2,n-2}$ where $t_{1-\alpha/2,n-2}$ is the $100(1 - \alpha/2)$th percentile of the $t$ distribution with $n - 2$ d.f. When the alternative hypothesis is the one-sided $H_1 : \{\sigma_1 > \sigma_2\}$, the null hypothesis is rejected if $PAN \geq t_{1-\alpha,n-2}$. When the alternative hypothesis is the one-sided $H_1 : \{\sigma_1 < \sigma_2\}$, the null hypothesis is rejected if $PAN \leq -t_{1-\alpha,n-2}$.

Marozzi (2011) showed that the resampling version of the Pan test should be preferred to the original one, and for this reason we consider here the permutation Pan test. To carry out such a test, we consider the resampling framework for scale testing described by Boos and Brownie (1989). We draw without replacement from $\mathbf{Y} = (X_{ji} - \overline{X}_j, i = 1, ..., n_j, j = 1, 2)$, where $\overline{X}_j$ is the mean of the $j$th sample, $B$ couples of samples of size $n_1$ and $n_2$, respectively, and compute the test statistic for each couple obtaining $PAN_b^*, b = 1, ..., B$. Let $PAN^o$ be the observed value of the test statistic, that is the $PAN$ statistic computed on $\mathbf{Y}$.

(i) One-sided upper-tail test. To test $H_0 : \{\sigma_1 = \sigma_2\}$ against $H_1 : \{\sigma_1 > \sigma_2\}$ at the $\alpha$ level of significance reject $H_0$ if

$$\frac{1}{B}\sum_{b=1}^{B} \mathbb{I}(PAN_b^* \geq PAN^o) < \alpha,$$

otherwise do not reject.

(ii) One-sided lower-tail test. To test $H_0 : \{\sigma_1 = \sigma_2\}$ against $H_1 : \{\sigma_1 < \sigma_2\}$ at the $\alpha$ level of significance reject $H_0$ if

$$\frac{1}{B}\sum_{b=1}^{B} \mathbb{I}(PAN_b^* \leq PAN^o) < \alpha,$$

otherwise do not reject.

(iii) Two-sided test. To test $H_0 : \{\sigma_1 = \sigma_2\}$ against $H_1 : \{\sigma_1 \neq \sigma_2\}$ at the $\alpha$ level of significance reject $H_0$ if

$$\frac{1}{B}\left[\sum_{b=1}^{B} \mathbb{I}(PAN_b^* \leq -|PAN^o|) + \sum_{b=1}^{B} \mathbb{I}(PAN_b^* \geq |PAN^o|)\right] < \alpha,$$

otherwise do not reject.

It should be noted that under $H_0$ $\mathbf{Y}$ elements are not exchangeable and so the permutation test does not yield an exact $p$-value. However the $p$-value converges to $\alpha$ as $min(n_1, n_2) \to \infty$ with $n_1/n_2 \to$ constant. For more details on the resampling approach for scale testing, see Boos and Brownie (1989, 2004), and Lim and Loh (1996).

To analyze the time in minutes to breakdown of an insulating fluid under elevated voltage stresses we should consider case (i), because the experimenter would like to test if the variability of times is significantly larger for the first power voltage (32 kV). To perform the test using $R$ load the file "pan_test.r" with the function pan(x1,x2,alt,B) where x1 and x2 are the samples, alt the type of alternative (alt="greater" or "less" or "two.sided" as usual) and B the number of permutations (default is B=10 000). The output of the function is the observed value $PAN^o$ and the $p$-value of the test. The $R$ code for performing the test using 10 000 permutations for estimating the $p$-value is:

```
> source("pan_test.r")
> #32k volt sample
> x1=c(.27,.40,.69,.79,2.75,3.91,9.88,13.95,15.93,27.80,53.24,82.85,
89.29,100.58,215.50)
> #36k volt sample
> x2=c(.35,.59,.96,.99,1.69,1.97,2.07,2.58,2.71,2.90,3.67,3.99,5.35,
13.77,25.50)
> pan.test=pan(x1,x2,alt="greater",B=10000)
```

The corresponding output is displayed by typing:

```
> pan.test$obs.value
[1] 4.115997
> pan.test$p.value
[1] 0
```

for the observed value of the test statistic and the $p$-value $\frac{1}{B} \sum_{b=1}^{B} \mathbb{I}(PAN_b^* \geq PAN^o)$ of the test, respectively.

The result shows that there is very strong practical evidence to reject the hypothesis of equal variability in favor of that of greater variability for the first voltage stress.

## 2.2.3    The permutation O'Brien test

In the previous subsection we described the permutation Pan test, which is a version of the Levene test particularly useful when the distributions underlying the samples of interest are heavy-tailed and/or highly skewed. Another version of the Levene test is particularly useful for symmetric and light-tailed distributions: this is the permutation

O'Brien test and it was considered for the first time by Marozzi (2011). The original O'Brien test is just the Student $t$ test on

$$R_{ji} = \frac{(n_j - 1.5)n_j(X_{ji} - \overline{X}_j)^2 - 0.5S_j^2(n_j - 1)}{(n_j - 1)(n_j - 2)},$$

where $S_j^2$ denotes the sample variance of $X_j, j = 1, 2$, therefore the O'Brien test is based on

$$OBR = \frac{\overline{R}_1 - \overline{R}_2}{\sqrt{\left(\frac{1}{n_1} + \frac{1}{n_2}\right) \frac{(n_1-1)W_1^2 + (n_2-1)W_2^2}{n-2}}},$$

where $\overline{R}_j$ and $W_j^2$ denote, respectively, the mean and the variance of $R_{ji}, i = 1, ..., n_j, j = 1, 2$. $H_0 : \{\sigma_1 = \sigma_2\}$ is rejected at the $\alpha$ level of significance in favor of the two-sided $H_1 : \{\sigma_1 \neq \sigma_2\}$ if $|OBR| \geq t_{1-\alpha/2, n-2}$. When the alternative hypothesis is the one-sided $H_1 : \{\sigma_1 > \sigma_2\}$, the null hypothesis is rejected if $OBR \geq t_{1-\alpha, n-2}$. When the alternative hypothesis is the one-sided $H_1 : \{\sigma_1 < \sigma_2\}$, the null hypothesis is rejected if $OBR \leq -t_{1-\alpha, n-2}$.

Marozzi (2011) showed that the resampling version of the O'Brien test should be preferred to the original one, and for this reason we consider here the permutation O'Brien test. To carry out such a test we draw without replacement from $Y$ $B$ couples of samples of size $n_1$ and $n_2$, respectively, and compute the test statistic for each couple obtaining $OBR_b^*, b = 1, ..., B$. Let $OBR^o$ be the observed value of the test statistic, that is the $OBR$ statistic computed on $Y$.

(i) One-sided upper-tail test. To test $H_0 : \{\sigma_1 = \sigma_2\}$ against $H_1 : \{\sigma_1 > \sigma_2\}$ at the $\alpha$ level of significance reject $H_0$ if

$$\frac{1}{B} \sum_{b=1}^{B} \mathbb{I}(OBR_b^* \geq OBR^o) < \alpha,$$

otherwise do not reject.

(ii) One-sided lower-tail test. To test $H_0 : \{\sigma_1 = \sigma_2\}$ against $H_1 : \{\sigma_1 < \sigma_2\}$ at the $\alpha$ level of significance reject $H_0$ if

$$\frac{1}{B} \sum_{b=1}^{B} \mathbb{I}(OBR_b^* \leq OBR^o) < \alpha,$$

otherwise do not reject.

Table 2.3   The FMR (in kJ/day) as a multiple of the basal metabolic rate of eight male and six female fulmars.

| | | | Male fulmars | | | | |
|---|---|---|---|---|---|---|---|
| 7.85 | 7.03 | 6.37 | 5.73 | 3.53 | 2.30 | 1.42 | 1.40 |

| | | | Female fulmars | | |
|---|---|---|---|---|---|
| 7.17 | 5.46 | 4.75 | 3.95 | 3.94 | 2.67 |

(iii)  Two-sided test. To test $H_0 : \{\sigma_1 = \sigma_2\}$ against $H_1 : \{\sigma_1 \neq \sigma_2\}$ at the $\alpha$ level of significance reject $H_0$ if

$$\frac{1}{B}\left[\sum_{b=1}^{B}\mathbb{I}(OBR_b^* \leq -|OBR^o|) + \sum_{b=1}^{B}\mathbb{I}(OBR_b^* \geq |OBR^o|)\right] < \alpha,$$

otherwise do not reject.

We use the permutation O'Brien test to address an ecological problem. Furness and Bryant (1996) studied the field metabolic rate (FMR; in kJ/day) of the Northern Fulmar (*Fulmarus glacialis*). The Northern Fulmar is one of the more abundant seabirds in the northern North Pacific and it is suspected to have rather lower energy expenditures than many other seabirds and therefore it is of interest to study its FMR (Table 2.3). We would like to test the null hypothesis of equal variability against the one-sided alternative that the variability of male FMR is greater than the variability of female FMR, this is case (i). To perform the test using $R$ load the file "obrien_test.r" with the function obrien(x1,x2,alt,B) where x1 and x2 are the samples, alt the type of alternative (alt="greater" or "less" or "two.sided" as usual) and B the number of permutations (default is B=10 000). The output of the function is the observed value $OBR^o$ and the $p$-value of the test. The $R$ code for performing the test using 10 000 permutations for estimating the $p$-value is:

```
> source("obrien_test.r")
> #Male subjects
> x1=c(7.85,7.03,6.37,5.73,3.53,2.3,1.42,1.4)
> #Female subjects
> x2=c(7.17,5.46,4.75,3.95,3.94,2.67)
> obrien.test=obrien(x1,x2,alt="greater",B=10000)
```

The corresponding output is displayed by typing:

```
> obrien.test$obs.value
[1] 1.898645
> obrien.test$p.value
[1] 0.0266
```

for the observed value of the test statistic and the $p$-value $\frac{1}{B} \sum_{b=1}^{B} \mathbb{I}(OBR_b^* \geq OBR^o)$ of the test, respectively.

The result shows that there is very strong practical evidence that the variability of male FMR is greater than the variability of female FMR.

A resampling test for variability comparison that has good performance both under symmetric normal- and light-tailed distributions, and under heavy-tailed and/or skewed ones has been proposed by Marozzi (2012a). The test is based on the combination of the Pan and the O'Brien test within the resampling framework. We do not consider here this combined test as well as those of Marozzi (2012b,d) because they are beyond the scope of this book. For the assessment of the two-sample problem within flexible adaptive two-stage designs, see Marozzi (2013b).

## 2.3    Jointly comparing central tendency and variability

The comparison of two samples is one of the most important problems in statistical testing. If it is assumed that parent population distributions may differ only in location, there are many parametric and nonparametric tests at our disposal (see Chapter 1). There are many tests also for the scale problem (see Section 2.2). It is well-known that under normal distributions the $t$ test and the $F$ test are the uniformly most powerful unbiased tests for the location and scale problem, respectively, at least for the one-sided alternative; and that the $t$ test is $\alpha$ robust for non-normal distributions (except for very heavy-tailed ones), whereas the $F$ test is non $\alpha$ robust (see, among others, Tiku *et al.*, 1986 and Wilcox, 2005). Even if the usual two-sample problem tests for a location change, in many biomedical situations the treatment can change location and scale simultaneously (see e.g., Muccioli *et al.*, 1996). The two-sample location-scale problem arises also in other fields such as climate dynamics (Yonetani and Gordon, 2001) and stock prices (Lunde and Timmermann, 2004).

Since in practice, normal data as well as either pure location or pure scale alternatives are not common, we need a test that works well for non-normal data in jointly testing for location and scale changes. Tukey (1959) emphasizes that a new test to be useful in practice should be quick and compact, in the sense that it should be easily used in practice without much effort by practitioners. If the procedure is quick and compact, then it can afford somewhat reduced power because it can be used so much more often as to more than compensate for its possible loss of power. Let us consider the two-stage testing scheme proposed by Manly and Francis (2002). In the first stage, an exact permutation test is carried out to determine whether there is evidence that the distributions underlying the samples to be compared are different. If the test is not significant, the procedure stops and it is concluded that there is no evidence of scale or location differences. If the test is significant, then a robust approximate permutation test for detecting location changes is carried out, together with a bootstrap procedure that indicates whether this test is reliable. A robust test for detecting scale changes is also performed. This testing scheme was found to control the type-one error rate even when the samples come from extreme distributions. The authors emphasized that this is achieved at the expense of often concluding that there is evidence for a difference between the populations under consideration without indicating the nature of the difference, and for this reason did not suggest a general use of the procedure. We emphasize another drawback of the procedure: it is neither quick nor compact in the sense of Tukey, since it is quite difficult to be applied by practitioners. For the same reason, in this subsection we are not interested in location-scale tests based on adaptive designs (for this approach see e.g., Büning and Thadewald, 2000 and Neuhäuser, 2001), or the tests proposed by Neuhäuser (2000), Büning (2002) and Murakami (2007). Note that, no formulae are available in the literature for the moments of some statistics necessary for applying the tests of Neuhäuser and Murakami (therefore to apply these tests one has to generate the permutation distribution of the statistics by simulation or complete enumeration, and compute the necessary moments).

In this subsection, we address the location-scale problem within the permutation/ rank framework. The best known and most used rank test is due to Lepage (1971). We consider also the Cucconi (1968) test, proposed earlier but not nearly as well known.

### 2.3.1   The Lepage test

We analyze the subset of the data obtained by Karpatkin *et al.* (1981) in their study of the effect of maternal steroid therapy on platelet counts of newborn babies, that was analyzed by Hollander and Wolfe (1999), see Table 2.4. Mothers of the babies were diagnosed with autoimmune thrombocytopenia purpura (ATP); those of the first group were treated with the corticosteroid prednisone and those of the second group were not. The aim of the treatment is to raise platelet counts of babies to safe levels during the delivery, in order to lower the possibility of intracranial hemorrhage for the baby. A woman with ATP produces antibodies to her own platelets and because of transplacental passage of antiplatelet antibodies during pregnancy, their babies are often born with low platelet counts, with a big concern that a vaginal delivery could cause intracranial hemorrhage. It is expected that the prednisone by crossing the placenta enters the baby's circulation and prevents splenic removal of those baby's platelets which are coated by the mother's antibodies. The primary aim of the study is whether or not predelivery maternal prednisone therapy increases newborn baby platelet counts. Even if the main statistical problem is detecting a possible difference

Table 2.4   Platelet counts (per cubic millimeter) of two groups of newborn babies.

| Case subjects | | | | | | | | | |
|---|---|---|---|---|---|---|---|---|---|
| 120 | 124 | 215 | 90 | 67 | 95 | 190 | 180 | 135 | 399 |
| Control subjects | | | | | | | | | |
| 12 | 20 | 112 | 32 | 60 | 40 | | | | |

in locations for the treated and not treated distribution populations, there is also some worries that the steroid therapy could rather largely increase the variability in newborn baby platelet counts. Therefore we regard changes in location and in scale as treatment effect.

In the literature, nonparametric tests for jointly detecting location and scale changes, that is to test

$$H_0 : \{\mu_1 = \mu_2 \cap \sigma_1 = \sigma_2\} \text{ against } H_1 : \{\mu_1 \neq \mu_2 \cup \sigma_1 \neq \sigma_2\},$$

are based on the combination of two tests, one for location and another for scale. Generally the combination is achieved through the sum of the squared standardized test statistics, this is just the case of the Lepage test, that is based on the following test statistic

$$LEP = \frac{(W - \mathbb{E}_0(W))^2}{\mathbb{V}_0(W)} + \frac{(AB - \mathbb{E}_0(AB))^2}{\mathbb{V}_0(AB)},$$

where $W$ is the Wilcoxon–Mann–Whitney statistic, and $AB$ the Ansari–Bradley one. $\mathbb{E}_0 (.)$ and $\mathbb{V}_0 (.)$ denote the expected value and variance of $W$ and $AB$ under $H_0$

$$\mathbb{E}_0 (W) = n_1 (n + 1) /2, \mathbb{V}_0 (W) = n_1 n_2 (n + 1) /12,$$

$$\mathbb{E}_0 (AB) = n_1 (n + 2) /4, \mathbb{V}_0 (AB) = n_1 n_2 (n + 2)(n - 2)/48/(n - 1)$$

$$\text{when } n \text{ is even,}$$

$$\mathbb{E}_0 (AB) = n_1 (n + 1)^2 /4/n, \mathbb{V}_0 (AB) = n_1 n_2 (n + 1)(n^2 + 3)/48/n^2$$

$$\text{when } n \text{ is odd.}$$

As for the Ansari–Bradley test for variability comparison, prior to the widespread of powerful and relatively cheap computers, to apply the Lepage test one had to look at the corresponding table of critical values or to consider its large sample approximation (Hollander and Wolfe, 1999). Moreover, in the presence of ties, the formula for computing it is rather tedious. Nowadays, using software like $R$, the test can be applied easily as a permutation test: first you compute the observed value $LEP^o$ of the Lepage statistic, next you simulate the permutation distribution $LEP_b^*, b = 1, ..., B$ of the test statistic by taking a large random sample of $B$ permutations, for example $B = 10\,000$, of the pooled sample $X = (X_1, X_2)$ and then you compute the corresponding $p$-value $\frac{1}{B} \sum_{b=1}^{B} \mathbb{I}(LEP_b^* \geq LEP^o)$.

To perform the test using $R$, load the file "lepage_test.r" with the function lepage(x1,x2,B) where x1 and x2 are the samples and B the number of permutations (default is B=10 000). Note that there is no distinction between alternative hypotheses since the alternative hypothesis is always the logic negation of the null hypothesis. The output of the function is the observed value $LEP^o$ and the $p$-value

$\frac{1}{B} \sum_{b=1}^{B} \mathbb{I}(LEP_b^* \geq LEP^o)$ of the test. The $R$ code for performing the test using $10\,000$ permutations for estimating the $p$-value is:

```
> source("lepage_test.r")
> #Case subjects
> x1=c(120,124,215,90,67,95,190,180,135,399)
> #Control subjects
> x2=c(12,20,112,32,60,40)
> lepage.test=lepage(x1,x2,B=10000)
```

The corresponding output is displayed by typing:

```
> lepage.test$obs.value
[1] 9.338375
> lepage.test$p.value
[1] 0.0038
```

for the observed value of the test statistic and the $p$-value of the test, respectively.

The Lepage test indicates to reject the null hypothesis of equal locations and scales, that is we detect a possible difference in locations and/or scales for the treated and not treated (with steroid therapy) distribution populations. We conclude that the two groups of newborn babies differ in terms of platelet counts.

## 2.3.2   The Cucconi test

The standard rank test for the two-sample location-scale problem is the Lepage test which is a combination of the Wilcoxon test for location and the Ansari–Bradley test for scale. Many rank tests have been proposed for the two-sample location-scale problem, nearly all of them are Lepage type tests, that is a combination of a location test and a scale test. Marozzi (2013a) reviewed and compared several two-sample location-scale tests of the Lepage type as well as the Cucconi (1968) test. The Cucconi test is not as familiar as other location-scale tests but it is of interest for several reasons. First, from a historical point of view it is of interest because it was proposed some years before the Lepage test. Secondly, it is of interest because it is not a combination of a test for location and a test for scale as the other location-scale tests. Thirdly, as shown by Marozzi (2009, 2013a) it compares favorably with Lepage type tests in terms of power and type-one error probability and very importantly it is easier to be computed because it requires only the ranks of one sample in the combined sample, whereas the other tests also require scores of various types as well as to permutationally estimate mean and variance of test statistics because their analytic formulae are not available.

After a period of more than 40 years, Marozzi (2009) put focus on the Cucconi test computing for the very first time a table of exact critical values (note that Cucconi

(1968) provided only the asymptotic critical values) and studying its type-one error probability and power. The Cucconi test is based on

$$CUC = \frac{U^2 + V^2 - 2\rho UV}{2(1 - \rho^2)},$$

where

$$U = \frac{6 \sum\limits_{i=1}^{n_1} R_{1i}^2 - n_1(n + 1)(2n + 1)}{\sqrt{n_1 n_2 (n + 1)(2n + 1)(8n + 11)/5}},$$

$$V = \frac{6 \sum\limits_{i=1}^{n_1} (n + 1 - R_{1i})^2 - n_1(n + 1)(2n + 1)}{\sqrt{n_1 n_2 (n + 1)(2n + 1)(8n + 11)/5}},$$

$n = n_1 + n_2$, $R_{ji}$ denotes the rank of $X_{ji}$ in the pooled sample $\mathbf{X} = (\mathbf{X}_1, \mathbf{X}_2)$ and

$$\rho = \frac{2(n^2 - 4)}{(2n + 1)(8n + 11)} - 1.$$

Note that $U$ is based on the squares of the ranks $R_{1i}$, while $V$ is based on the squares of the contrary ranks $(n + 1 - R_{1i})$ of the first sample. In the absence of ties $U' = -U$ and $V' = -V$ where $U'$ and $V'$ are $U$ and $V$ computed on the second sample and then it does not matter if one acts on the first or second sample to compute the $CUC$ statistic. In the presence of ties, the $CUC$ statistic computed on the first and second sample gives rise to very slightly different values. This aspect had been already noted by Cucconi (1968) and recently by Neuhäuser (2012). For practical purposes, in this case we suggest to average the two values of the $CUC$ statistic, as is done by the R function which performs the Cucconi test that will be considered later on. Under $H_0 : \{\mu_1 = \mu_2 \cap \sigma_1 = \sigma_2\}$, $\mathbb{E}(U) = \mathbb{E}(V) = 0$ and $\mathbb{V}(U) = \mathbb{V}(V) = 1$ because

$$\mathbb{E}\left( \sum_{i=1}^{n_1} R_{1i}^2 \right) = n_1(n + 1)(2n + 1)/6$$

and

$$\mathbb{V}\left( \sum_{i=1}^{n_1} R_{1i}^2 \right) = n_1 n_2 (n + 1)(2n + 1)(8n + 11)/180.$$

Of course,

$$\mathbb{E}\left( \sum_{i=1}^{n_1} (n + 1 - R_{1i})^2 \right) = \mathbb{E}\left( \sum_{i=1}^{n_1} R_{1i}^2 \right), \mathbb{V}\left( \sum_{i=1}^{n_1} (n + 1 - R_{1i})^2 \right) = \mathbb{V}\left( \sum_{i=1}^{n_1} R_{1i}^2 \right),$$

and $U$ and $V$ are negatively dependent. More precisely, $\mathbb{C}or(U, V)$ takes values in the interval $[-1, -7/8)$, in fact

$$\mathbb{C}or(U, V) = \mathbb{C}ov(U, V) = \frac{2(n^2 - 4)}{(2n + 1)(8n + 11)} - 1 = \rho,$$

the minimum $-1$ occurs when $n = 2$, while the supremum is reached when $n$ tends to infinity

$$\lim_{n \to \infty} \rho = \lim_{n \to \infty} \frac{2(n^2 - 4)}{(2n + 1)(8n + 11)} - 1 = -\frac{7}{8} = \rho_0.$$

Under $H_0$, $(U, V)$ is centered on $(0,0)$, whereas it is not under $H_1$. When $\mu_1 \neq \mu_2$ and $\sigma_1 = \sigma_2$, $\sum_{i=1}^{n_1} R_{1i}^2$ tends to be greater (less) than $n_1(n + 1)(2n + 1)/6$ when $\mu_1 > \mu_2$ ($\mu_1 < \mu_2$) and so $U$ tends to be greater (less) than 0 while $V$ tends to be less (greater) than 0. When $\mu_1 = \mu_2$ and $\sigma_1 > \sigma_2$, the ranks of the elements of the first sample tend to be the extreme elements of the sequence $1, 2, \ldots, n$, while when $\sigma_1 < \sigma_2$ they tend to be in the middle. In the first (second) case, $\sum_{i=1}^{n_1} R_{1i}^2$ tends to be greater (less) than $n_1(n + 1)(2n + 1)/6$ and then $U$ tends to be greater (less) than 0 and so does $V$. When $\mu_1 \neq \mu_2$ and $\sigma_1 \neq \sigma_2$, $(U, V)$ is once again non centered on $(0,0)$, for example when $\mu_1 > \mu_2$ and $\sigma_1 < \sigma_2$, $\mathbb{E}(U)$ may be close to 0 but $\mathbb{E}(V)$ tends to be less than 0. With standard asymptotics it may be proved that if $n_1, n_2 \to \infty$ and $n_1/n \to \lambda \in (0, 1)$ then $\Pr(U \leq u) \to \Phi(u)$, where $\Phi$ is the standard normal distribution function. An analogous result applies for $V$. Moreover if $n_1, n_2 \to \infty$ and $n_1/n \to \lambda \in (0, 1)$ then

$$\Pr(U \leq u, V \leq v) \to \int_{-\infty}^{u} \int_{-\infty}^{v} \frac{1}{2\pi \sqrt{1 - \rho_0^2}} \exp\left(-\frac{q^2 + r^2 - 2\rho_0 qr}{2(1 - \rho_0^2)}\right) dq dr.$$

The points $(u, v)$ within acceptance region are close to $(0, 0)$, that is satisfy

$$\frac{1}{2\pi \sqrt{1 - \rho_0^2}} \exp\left(-\frac{u^2 + v^2 - 2\rho_0 uv}{2(1 - \rho_0^2)}\right) \geq k,$$

where the constant $k$ is chosen so that the type-one error rate is $\alpha$. Let $k = \alpha\left(2\pi \sqrt{1 - \rho_0^2}\right)^{-1}$, then it follows that $H_0$ should be accepted if the point $(u, v)$ is such that $\frac{u^2 + v^2 - 2\rho_0 uv}{2(1 - \rho_0^2)} < -\ln \alpha$. It should be noted that the acceptance region $E$ of the test is the set of points $(u, v)$ inside the ellipse

$$u^2 + v^2 - 2\rho_0 uv = -2\left(1 - \rho_0^2\right) \ln \alpha.$$

Since it is

$$\int\int_E \frac{1}{2\pi\sqrt{1-\rho_0^2}} \exp\left(-\frac{q^2+r^2-2\rho_0 qr}{2\left(1-\rho_0^2\right)}\right) dqdr = 1-\alpha,$$

the size of test is just $\alpha$. In practice, unless you have large samples, $\rho_0$ should be replaced by $\rho$. Cucconi (1968) observed that the rate of convergence to the normal was very good as soon as $n_1, n_2 > 6$ with $n_1$ and $n_2$ not very different. However this is not of much interest for us since we consider here the permutation version of the Cucconi test. This test can be easily performed using $R$. We show this by reanalyzing the data considered in the previous subsection about the effect of maternal steroid therapy on platelet counts of newborn babies. Marozzi (2012c) proposed a modified version of the Cucconi test which is not considered here because it is beyond the scope of this book. It is obtained within the nonparametric combination framework, see Section 1.2.2. This framework has been considered also in addressing location problems (Marozzi, 2002, 2003, 2004a,b, 2007).

To perform the permutation version of the Cucconi test, first compute the observed value $CUC^o$ of the Cucconi statistic, next simulate the permutation distribution $CUC_b^*, b = 1, ..., B$ of the test statistic by taking a large random sample of $B$ permutations, for example $B = 10\,000$, of the pooled sample $X = (X_1, X_2)$ and then compute the corresponding $p$-value $\frac{1}{B}\sum_{b=1}^B \mathbb{I}(CUC_b^* \geq CUC^o)$. To perform the test using $R$ load the file "cucconi_test.r" with the function cucconi(x1,x2,B) where x1 and x2 are the samples and B the number of permutations (default is B=10 000). Note that as for the Lepage test, there is no distinction between alternative hypotheses since the alternative hypothesis is always the logic negation of the null hypothesis. The output of the function is the observed value $CUC^o$ and the $p$-value $\frac{1}{B}\sum_{b=1}^B \mathbb{I}(CUC_b^* \geq CUC^o)$ of the test. The $R$ code for performing the test using 10 000 permutations for estimating the $p$-value is:

```
> source("cucconi_test.r")
> #Case subjects
> x1=c(120,124,215,90,67,95,190,180,135,399)
> #Control subjects
> x2=c(12,20,112,32,60,40)
> cucconi.test=cucconi(x1,x2,B=10000)
```

The corresponding output is displayed by typing:

```
> cucconi.test$obs.value
[1] 4.692997
> cucconi.test$p.value
[1] 0.0031
```

for the observed value of the test statistic and the $p$-value of the test, respectively.

The Cucconi test is concordant with the Lepage test (see Section 2.3.1) significantly detecting a possible difference in locations and/or scales for the treated and not treated (with steroid therapy) distribution populations. We conclude that the two groups of newborn babies differ in terms of platelet counts.

## 2.4    Comparing distributions

The problem of comparing two samples has been addressed in Section 2.3 by jointly testing for location and scale changes. In this subsection the problem is addressed from the most general point of view, in fact we are interested in assessing whether there are any differences between the distributions of populations underlying the samples. The tests presented here are sensitive to all possible types of differences between the two distribution functions associated with the two populations, including differences in shape, and thus in kurtosis and/or skewness. Therefore, when the researcher suspects that differences in shape as well as differences in location and/or scale (between the two distributions) are possible, it is suggested to use the tests presented in the following subsections.

### 2.4.1    The Kolmogorov–Smirnov test

Household expenditures (in Hong Kong dollars) of 20 single men and 20 single women including fuel and light are reported in Table 2.5. This real data example is taken from Hand *et al.* (1994, p. 44). We suspect that the distributions of the data in the two groups have different locations as well as different scales. Moreover it is suspected that the distribution functions associated with the populations behind the samples may differ also in shape. Therefore this is a situation where a test for the most general alternative hypothesis is more suitable than a test for the location-scale problem (like the Lepage test or the Cucconi test, see Section 2.3).

A very familiar test for the general two-sample problem is the Kolmogorov–Smirnov test which requires Assumptions A1 and A2 and it is based on the differences between the empirical distribution functions (EDFs) of the two samples. If the random variables $X_1$ and $X_2$ underlying the samples are continuous, the test is exact. In the discrete case, or when in practice you have ties, the test is conservative (its significance

Table 2.5    Household expenditures (in Honk Kong dollars) of a group of men and a group of women.

| | | | | Men | | | | | |
|---|---|---|---|---|---|---|---|---|---|
| 497 | 839 | 798 | 892 | 1585 | 755 | 388 | 617 | 248 | 1641 |
| 1180 | 619 | 253 | 661 | 1981 | 1746 | 1865 | 238 | 1199 | 1524 |

| | | | | Women | | | | | |
|---|---|---|---|---|---|---|---|---|---|
| 820 | 184 | 921 | 488 | 721 | 614 | 801 | 396 | 864 | 845 |
| 404 | 781 | 457 | 1029 | 1047 | 552 | 718 | 495 | 382 | 1090 |

level is less than or equal to the nominal $\alpha$ level). Let $F_1$ and $F_2$ be the distribution functions corresponding to populations 1 and 2, respectively. Let $EDF_1$ and $EDF_2$ denote the EDFs of the $X_1$ and $X_2$ samples

$$EDF_j(x) = \frac{1}{n_j} \sum_{i=1}^{n_j} \mathbb{I}(X_{1i} \leq x), \text{ for } j = 1, 2.$$

The test statistic differs according to the system of hypotheses (two- or one-sided) of interest. In the two-sided case, the null hypothesis is

$$H_0 : \{F_1(x) = F_2(x), \forall x \in \mathcal{R}\}$$

and the alternative hypothesis is

$$H_1 : \{F_1(x) \neq F_2(x) \text{ for at least one } x \in \mathcal{R}\}.$$

The test statistic is the maximum vertical distance between the two EDFs

$$KS = \sup_{x \in \mathcal{R}} |EDF_1(x) - EDF_2(x)|.$$

In the one-sided case, we distinguish two situations. The first situation is when the first variable is stochastically larger than the second variable in the alternative hypothesis. In this case we test

$$H_0 : \{F_1(x) \leq F_2(x), \forall x \in \mathcal{R}\}$$

against

$$H_1 : \{F_1(x) \geq F_2(x) \text{ for at least one } x \in \mathcal{R}\}.$$

The test statistic is the maximum vertical distance attained by $EDF_1$ above $EDF_2$

$$KS^+ = \sup_{x \in \mathcal{R}}(EDF_1(x) - EDF_2(x)).$$

Note that when $X_1$ is stochastically larger than $X_2$, the distribution function of $X_1$ lies below and hence to the right of that of $X_2$.

The second situation is when the first variable is stochastically smaller than the second variable in the alternative hypothesis. In this case we test

$$H_0 : \{F_1(x) \geq F_2(x), \forall x \in \mathcal{R}\}$$

against

$$H_1 : \{F_1(x) \leq F_2(x) \text{ for at least one } x \in \mathcal{R}\}.$$

The test statistic is the maximum vertical distance attained by $EDF_2$ above $EDF_1$

$$KS^- = \sup_{x \in \mathcal{R}}(EDF_2(x) - EDF_1(x)).$$

Note that when $X_1$ is stochastically smaller than $X_2$, the distribution function of $X_1$ lies above and hence to the left of that of $X_2$. Note that the Kolmogorov–Smirnov test is always significant for large values of the test statistic.

In practice ties may occur. Since the EDF is well defined also in the presence of ties, the Kolmogorov–Smirnov statistics do not need any adjustments. The unique shortcoming is that the test becomes conservative.

The test has a natural motivation. Since the EDFs are estimators of $F_1$ and $F_2$, therefore the Kolmogorov–Smirnov statistics may be seen as estimators of the maximum vertical difference between $F_1$ and $F_2$, which is 0 when the null hypothesis is true. Large values of the appropriate statistic are evidence in favor of the specified alternative hypothesis. It is important to note that the Kolmogorov–Smirnov test, under Assumptions A1 and A2 is consistent against any differences between $F_1$ and $F_2$. This protection against all possible difference types is traded for a possible reduced power against specific alternatives (like pure location, pure scale or location-scale alternatives; Marozzi, 2009, 2013a). The exact null distribution of the Kolmogorov–Smirnov statistic is found by computing the test statistic for all orderings of $X_1$ and $X_2$, which are equally likely under the null hypothesis. Note that to compute in practice the appropriate Kolmogorov–Smirnov statistic, we can compute the vertical differences between the EDFs at the ordered sample observations in the combined sample $X_{(1)} \leq \ldots \leq X_{(n)}$ rather than for all $x \in \mathcal{R}$:

$$KS = \max_{i=1,\ldots,n} |EDF_1\left(X_{(i)}\right) - EDF_2\left(X_{(i)}\right)|$$

$$KS^+ = \max_{i=1,\ldots,n} \left(EDF_1\left(X_{(i)}\right) - EDF_2\left(X_{(i)}\right)\right)$$

$$KS^- = \max_{i=1,\ldots,n} \left(EDF_2\left(X_{(i)}\right) - EDF_1\left(X_{(i)}\right)\right).$$

The reason is that the EDFs are step functions that change values only at the observed sample values.

Before the spread of fast and cheap computers, to perform the test one had to look at tables of critical values or to consider some asymptotic approximations. The basic package of $R$ contains the function ks.test which computes both the appropriate Kolmogorov–Smirnov statistic and the $p$-value. In the two-sided case, if $n_1 n_2 < 10\,000$ and in the absence of ties the exact $p$-value is computed. In the other cases, asymptotic distributions are used. Note that they may be rather inaccurate for small samples. Note that in the presence of ties $R$ always generates a warning, since continuous distributions (Assumption A1) do not generate them. Data about household expenditures can be analyzed in $R$ using the following code:

```
> #men expenditures X1
> x1=c(497,839,798,892,1585,755,388,617,248,1641,1180,619,253,661,
1981,1746,1865,238,1199,1524)
> #women expenditures X2
> x2=c(820,184,921,488,721,614,801,396,864,845,404,781,457,1029,
1047,552,718,495,382,1090)
```

```
> ks.test(x1,x2,alternative="two.sided")
###################################################################
# Two-sample Kolmogorov--Smirnov test
#
#data: x1 and x2
#D = 0.4, p-value = 0.08106
#alternative hypothesis: two-sided
###################################################################
```

We considered the two-sided alternative because we would like to find evidence of whether the distribution functions of the populations underlying the samples are identical or not against all types of differences that may exist between the two distribution functions. The $p$-value of the test is 0.081 and indicates that there is moderate evidence to reject the hypothesis of equal distribution functions.

To obtain the observed value of the Kolmogorov–Smirnov statistic write:

```
> ks.test(x1,x2,alternative="two.sided")$statistic
D 0.4
```

and to obtain the $p$-value write:

```
> ks.test(x1,x2,alternative="two.sided")$p.value
[1] 0.08105771
```

If you suspect that male expenditures tend to be larger than female expenditures, you suspect that the expenditure distribution function for male tends to be smaller than the expenditure distribution function for female ($F_1$ lies below and to the right of $F_2$) therefore you should consider the first one-sided situation and test $H_0 : \{F_1(x) \leq F_2(x), \forall x \in \mathcal{R}\}$ against $H_1 : \{F_1(x) \geq F_2(x)$ for at least one $x \in \mathcal{R}\}$. In $R$ you have just to replace alternative="two.sided" with alternative="less". If you suspect that male expenditures tend to be smaller than female expenditures, you suspect that the expenditure distribution function for male tends to be larger than the expenditure distribution function for female ($F_1$ lies above and to the left of $F_2$) therefore you should consider the second one-sided situation and test $H_0 : \{F_1(x) \geq F_2(x), \forall x \in \mathcal{R}\}$ against $H_1 : \{F_1(x) \leq F_2(x)$ for at least one $x \in \mathcal{R}\}$. In $R$ you have just to replace alternative="two.sided" with alternative="larger". Note that when using t.test or wilcox.test functions, the one-sided alternatives are reversed (for example if you suspect that $X_1$ is stochastically larger than $X_2$ you have to use alternative="larger").

### 2.4.2    The Cramér–von Mises test

This subsection deals with another very familiar test for the general two-sample problem: the Cramer–von Mises test. Some people prefer it to the

Kolmogorov–Smirnov test because it seems to make more effective use of the data. Nevertheless the Cramer–von Mises test is slightly more difficult to be computed than the Kolmogorov–Smirnov test. It is important to note that the Cramer–von Mises test is two-sided only and then it can be used to test

$$H_0 : \{F_1(x) = F_2(x), \forall x \in \mathcal{R}\}$$

against

$$H_1 : \{F_1(x) \neq F_2(x) \text{ for at least one } x \in \mathcal{R}\}.$$

The Cramer–von Mises test requires the same assumptions of the Kolmogorov–Smirnov test, namely, A1 and A2. Several versions of the Cramer–von Mises test have been proposed in the literature. We consider here the L1 (taxicab distance) version studied by Xiao *et al.* (2006). This version is almost as powerful as the more familiar L2 (Euclidean distance) version but it is much more computationally efficient. The test is based on the following statistic

$$CVM = \frac{(n_1 n_2)^{1/2}}{n^{3/2}} \left( \sum_{i=1}^{n_1} |EDF_1(X_{1i}) - EDF_2(X_{1i})| \right.$$
$$\left. + \sum_{i=1}^{n_2} |EDF_1(X_{2i}) - EDF_2(X_{2i})| \right).$$

Large values of the *CVM* statistic are evidence against the equality between the two distribution functions associated with the populations behind the samples. The rationale of the Cramer–von Mises test is similar to that of the Kolmogorov–Smirnov test. $EDF_1$ and $EDF_2$ are estimates of $F_1$ and $F_2$ and the *CVM* statistic is a function of certain L1 distances between the two EDFs. Under the null hypothesis of equality between $F_1$ and $F_2$ it is suspected that the differences between $EDF_1$ and $EDF_2$ are not null only by chance. Under the alternative hypothesis, the more different $F_1$ and $F_2$, the larger the *CVM* statistic.

To find the exact null distribution of the Cramer–von Mises statistic, as for the Kolmogorov–Smirnov statistic you should compute *CVM* for all ordered arrangements of the combined sample, that are equally likely under the null hypothesis. Tables of critical values as well as asymptotic approximations are available. We do not consider them because we will use *R* to compute the *CVM* statistic as well as the corresponding exact *p*-value.

We show how to perform the Cramer–von Mises test using *R* by reanalyzing the data considered in Section 2.4.1 about household expenditures of a group of men and a group of women. The test can be performed in *R* using the CvM2SL1Test additional package that can be downloaded from http://cran.r-project.org/web/packages/CvM2SL1Test/index.html.

The analysis of household expenditures data can be performed running the following *R* code:

```
> #load the additional package
> library(CvM2SL1Test)
> #men expenditures X1
> x1=c(497,839,798,892,1585,755,388,617,248,1641,1180,619,253,661,
1981,1746,1865,238,1199,1524)
> #women expenditures X2
> x2=c(820,184,921,488,721,614,801,396,864,845,404,781,457,1029,
1047,552,718,495,382,1090)
> #this is the CVM statistic
> cvmtsl1.test(x1,x2)
[1] 0.466436
> n1=length(x1)
> n2=length(x2)
> #this is the exact p-value
> cvmtsl1.pval(cvmtsl1.test(x1,x2),n1,n2)
[1] 0.1376554
```

The *p*-value of the test is 0.138 and indicates that there is little evidence to reject the null hypothesis of equal distribution functions. Therefore looking at the Cramer–von Mises test results we conclude that the distribution functions associated with the populations underlying men and women household expenditure samples are not significantly different. If you are interested in the L2 version of the Cramer–von Mises test you should download the `CvM2SL2Test` additional package and run analogous code as for the L1 version of the test: `cvmts.test(x1,x2)` to compute the statistic and `cvmts.pval(cvmts.test(x1,x2),n1,n2)` to compute the corresponding *p*-value.

# References

Balakrishnan, N. and Ma, C.W. (1990) A comparative study of various tests for the equality of two population variances. Journal of Statistical Computation and Simulation, 35, 41–89.

Boos, D.D. and Brownie, C. (1989) Bootstrap methods for testing homogeneity of variances. Technometrics, 31, 69–82.

Boos, D.D. and Brownie, C. (2004) Comparing variances and other measures of dispersion. Statistical Science, 19, 4, 571–578.

Box, G.E.P (1953) Non-normality and tests on variances. Biometrika, 40, 318–335.

Brown, M.B. and Forsythe, A.B. (1974) Robust tests for the equality of variances. Journal of the American Statistical Association, 69, 364–367.

Büning, H. (2002) Robustness and power of modified Lepage, Kolmogorov-Smirnov and Cramèr-Von Mises two-sample tests. Journal of Applied Statistics, 29, 907–924.

Büning, H. and Thadewald, T. (2000) An adaptive two-sample location-scale test of Lepage type for Symmetric distributions. Journal of Statistical Computation and Simulation, 65, 287–310.

Conover, W.J., Johnson, M.E. and Johnson, M.M. (1981) A comparative study of tests for homogeneity of variances, with applications to the outer continental shelf bidding data. Technometrics, 23, 351–361.

Cucconi, O. (1968) Un nuovo test non parametrico per il confronto tra due gruppi campionari. Giornale degli Economisti, 27, 225–248.

Fairfull, R.W., Crober, D.C. and Gowe, R.S. (1985) Effects of comb dubbing on the performance of laying stocks. Poultry Science, 64, 434–439.

Furness, R.W., and Bryant, D.M. (1996) Effect of wind on field metabolic rates of breeding northern fulmars. Ecology, 77, 1181–1188.

Games, P.A., Winkler, H.B. and Probert, D.A. (1972) Robust tests for homogeneity of variance. Educational and Psychological Measurement, 32, 887–909.

Geng, S., Wang, W.J. and Miller, C. (1979) Small sample size comparisons of tests for homogeneity of variances by Monte-Carlo. Communications in Statistics – Simulation and Computation, 8, 379–389.

Gibbons, J.D. and Chakraborti, S. (2003) Nonparametric Statistical Inference, 4th edn. CRC Press.

Good, P. (2000) Permutation Tests, a Practical Guide to Resampling Methods for Testing Hypotheses, 2nd edn. Springer-Verlag.

Hall, I.J. (1972) Some comparisons of tests for equality of variances. Journal of Statistical Computation and Simulation, 1, 183–194.

Hand, D.J., Daly, F., Lunn, A.D., McConway, K.J. and Ostrowski, E. (1994) A Handbook of Small Data Sets. Chapman and Hall.

Hollander, M. and Wolfe, D.A. (1999) Nonparametric Statistical Methods, 2nd edn. John Wiley & Sons, Ltd.

Karpatkin, M, Porges, R.F. and Karpatkin, S. (1981) Platelet counts in infants of women with autoimmune thrombocytopenia: effects of steroid administration to the mother. New England Journal of Medicine, 305, 936–939.

Keselman, H.J., Games, P.A. and Clinch, J.J. (1979) Tests for homogeneity of variance, Communications in Statistics – Simulation and Computation, 8, 113–119.

Lepage, Y. (1971) A combination of Wilcoxon's and Ansari–Bradley's statistics, Biometrika. 58, 213–217.

Lim, T.S. and Loh, W.Y. (1996) A comparison of tests of equality of variances. Computational Statistics and Data Analysis, 22, 287–301.

Lunde, A. and Timmermann, A. (2004) Duration dependence in stock prices: an analysis of bull and bear markets. Journal of Business and Economic Statistics, 22, 253–273.

Manly, B.F.J. and Francis, R.I.C.C. (2002) Testing for mean and variance differences with samples from distributions that may be non-normal with unequal variances. Journal of Statistical Computation and Simulation, 72, 633–646.

Marozzi, M. (2002) Some notes on nonparametric inferences and permutation tests. Metron, 60, 139–151.

Marozzi, M. (2003) Applications in business, medical and industrial statistics of bi-aspect nonparametric tests for location problems. Statistical Methods and Applications, 12, 187–194.

Marozzi, M. (2004a) A bi-aspect nonparametric test for the two-sample location problem. Computational Statistics and Data Analysis, 44, 639–648.

Marozzi, M. (2004b) A bi-aspect nonparametric test for the multi-sample location problem. Computational Statistics and Data Analysis, 46, 81–92.

Marozzi, M. (2007) Multivariate tri-aspect non-parametric testing. Journal of Nonparametric Statistics, 19, 269–282.

Marozzi, M. (2009) Some notes on the location-scale cucconi test. Journal of Nonparametric Statistics, 21, 5, 629–647.

Marozzi, M. (2011) Levene type tests for the ratio of two scales. Journal of Statistical Computation and Simulation, 81, 815–826.

Marozzi, M. (2012a) A distribution free test for the equality of scales. Communication in Statistics – Simulation and Computation, 41, 878–889.

Marozzi, M. (2012b) A modified Hall-Padmanabhan test for the homogeneity of scales. Communication in Statistics – Theory and Methods, 41, 3068–3078.

Marozzi, M. (2012c) A modified Cucconi test for location and scale change alternatives. Colombian Journal of Statistics, 35, 369–382.

Marozzi, M. (2012d) A combined test for differences in scale based on the interquantile range. Statistical Papers, 53, 61–72.

Marozzi, M. (2013a) Nonparametric simultaneous tests for location and scale testing: a comparison of several methods. Communication in Statistics – Simulation and Computation, 42, 1298–1317.

Marozzi, M. (2013b) Adaptive choice of scale tests in flexible two-stage designs with applications in experimental ecology and clinical trials. Journal of Applied Statistics, 40, 747–762.

Murakami, H. (2007) Lepage type statistic based on the modified Baumgartner statistic. Computational Statistics and Data Analysis, 51, 5061–5067.

Miller, R.G. Jr. (1968) Jackknifing variances. Annals of Mathematical Statistics, 39, 567–582.

Muccioli, C., Belford, R., Podgor, M., Sampaio, P., de Smet, M. and Nussenblatt, R. (1996) The diagnosis of intraocular inflammation and cytomegalovirus retinitis in HIV-infected patients by laser flare photometry. Ocular Immunology and Inflammation, 4, 75–81.

Nair, V.N. (1984) On the behaviour of some estimators from probability plots. Journal of the American Statistical Association, 79, 823–830.

Nair, V.N. and Pregibon, D. (1988) Analyzing dispersion effects from replicated factorial experiments. Technometrics, 30, 247–257.

Neuhäuser, M. (2000) An exact two-sample test based on the Baumgartner-Weiss-Schindler statistic and a modification of Lepage's test. Communication in Statistics – Theory and Methods, 29, 67–78.

Neuhäuser, M. (2001) An adaptive location-scale test. Biometrical Journal, 43, 809–819.

Neuhäuser M. (2012) Nonparametric Statstical Tests: a Computational Approach. CRC Press.

O'Brien, R.G. (1979) A general ANOVA method for robust tests of additive models for variances. Journal of the American Statistical Association, 74, 877–880.

Olejnik, S.F. and Algina, J. (1988) Tests of variance equality when distributions differ in form and location. Educational and Psychological Measurement, 48, 317–329.

Pan, G. (1999) On a Levene type test for equality of two variances. Journal of Statistical Computation and Simulation, 63, 59–71.

Tan, W.Y. (1982) Sampling distributions and robustness of t, F and variance-ratio in two samples, and ANOVA models with respect to departure from normality. Communications in Statistics – Theory and Methods, 11, 2485–2511.

Tiku, M.L., Tan, W.Y. and Balakrishnan, N. (1986) Robust Inference. Marcel Dekker.

Tukey, J.W. (1959) A quick, compact, two-sample test to Duckworth's specifications. Technometrics, 1, 32–48.

Wilcox, R.R. (2005) Introduction to Robust Estimation and Hypothesis Testing. Academic Press.

Xiao, Y., Gordon, A. and Yakovlev, A. (2006) The L1-version of the Cramer-von Mises test for two-sample comparisons in microarray data analysis. EURASIP Journal on Bioinformatics and Systems Biology, 2006, 1–9.

Yonetani, T. and Gordon H.B. (2001) Abrupt changes as indicators of decadal climate variability. Climate Dynamics, 17, 249–258.

# 3

# Comparing more than two samples

## 3.1 Introduction

The problems presented in this chapter can be considered a generalization of the one-sample and two-sample location problems, in the sense that more than two samples are compared. The primary interest here consists in comparing the location of three or more samples. Under the null hypothesis the sample data are supposed to come from the same population. Under the alternative hypothesis data are supposed to come from different populations (or groups). In the analysis of variance problem (ANOVA) the variability of data is broken down into two components: (1) between groups variability, due to the different locations of the populations; (2) within groups variability, due to the variability in the populations. The focus of the analysis is on the significance of the first type of variability.

When the populations are defined according to different levels of one factor (or different treatments) we are in the so called one-way ANOVA layout. In this case the dataset is represented by $\{X_{ji}; i = 1, \ldots, n_j; j = 1, \ldots, C\}$, where $n_j$ is the sample size of the $j$th group and $C > 2$ is the number of groups or samples. The factor levels are represented by the integers $\{1, \ldots, C\}$. In the presence of two factors, the problem takes the name of two-way ANOVA. The dataset is $\{X_{jri}; i = 1, \ldots, n_{jr}; j = 1, \ldots, C_1; r = 1, \ldots, C_2\}$, where $n_{jr}$ is the number of sample observations in the group where the first factor is at level $j$ and the second factor at level $r$, and $C_1$ and $C_2$ are the numbers of possible levels for the first and second factor, respectively. Section 3.2 is dedicated to testing problems related to the one-way ANOVA layout. In Section 3.3 we deal with the two-way ANOVA layout problem. The particular case of multisample problems where the interest is focused on pairwise comparisons between groups is discussed in

*Nonparametric Hypothesis Testing: Rank and Permutation Methods with Applications in R*,
First Edition. Stefano Bonnini, Livio Corain, Marco Marozzi and Luigi Salmaso.
© 2014 John Wiley & Sons, Ltd. Published 2014 by John Wiley & Sons, Ltd.
Companion website: http://www.wiley.com/go/hypothesis_testing

Section 3.4. Section 3.5 concerns the multivariate extension of multisample problems. A typical representation of data in the presence of one factor in the multivariate case is $\{X_{jih}; \ i = 1, \ldots, n_j; \ j = 1, \ldots, C; \ h = 1, \ldots, q\}$ where as usual $n_j$ is the size of the $j$th sample, $C$ is the number of groups and $h$ denotes the component variable of the $q$-variate response, with $q \geq 2$.

## 3.2    One-way ANOVA layout

Let us consider statistical analyses where the main interest is focused on the locations of three or more populations. In this case we are in the presence of $C$ ($\geq 3$) random samples, each coming from one of the $C$ compared populations. The null hypothesis of interest is that of no difference in location, that is the $C$ samples can be treated as a single pooled sample from one population (Hollander and Wolfe, 1999). In this framework we are in the well-known one-way ANOVA design. In other words, the number $C$ of populations represents the number of levels of the symbolic treatment, and for the units in the $j$th group ($j = 1, \ldots, C$) the treatment is at level $j$. The only assumptions are: the $C$ populations are independent, data from population $j$ come from a continuous distribution with cumulative distribution function (CDF) $F_j(x)$ and the treatment produces an effect only on the location of the populations. Formally $F_j(x) = F(x + \theta_j)$ for every $x$ and for $j = 1, \ldots, C$, where $F$ is the CDF of a continuous distribution common to all populations and $\theta_j$ is the treatment effect for the $j$th population. Assuming that the $i$th observation of the $j$th sample $X_{ji}$ is a realization of the random variable $Z_{ji}$, according to the fixed effects additive model, we may formally write

$$Z_{ji} = \mu + \theta_j + \epsilon_{ji},$$

with $i = 1, \ldots, n_j$ and $j = 1, \ldots, C$, where $\mu$ is a population constant (e.g., unknown mean or median), $\theta_j$ is the unknown effect of the $j$th treatment and $\epsilon_{ji}$ are exchangeable random errors with zero mean or median and $\sigma$ standard deviation which is assumed to be invariant with respect to groups. Note that in this model the responses are assumed to be homoscedastic (equal variances) and the scale coefficients are assumed not to be affected by the treatment levels, in particular in the alternative hypothesis (Pesarin and Salmaso, 2010).

   Thus, the goal of the analysis in this framework consists of testing for differences among the treatment effects. Such a situation is very common in several real problems. A typical example is related to medical studies where we wish to compare $C$ different therapies for a given disease. Another interesting example concerns botany, when for example $C$ different fertilizers are studied, and the effects on height (in centimeters) of plants are compared. Of course the term treatment in general does not always correspond to the concept of 'cure'. We may consider for instance an industrial experiment in which one wishes to test the effect of the diameter (in millimeters) of a steel bar on the yield strength. Here the treatment levels correspond to $C$ different diameter sizes.

Formally the null hypothesis can be written as follows:

$$H_0 : \theta_1 = \theta_2 = \cdots = \theta_C.$$

This hypothesis asserts that the underlying distributions $F_1, \ldots, F_C$ are identical, that is the data of the $C$ samples come from one population. According to the alternative hypothesis, at least two of the treatment effects are not equal:

$$H_1 : \theta_1, \theta_2, \ldots, \theta_C \text{ are not equal,}$$

that is there exists at least one couple $(j, r)$ such that $\theta_j \neq \theta_r$ or equivalently $F_j \neq F_r$.

## 3.2.1   The Kruskal–Wallis test

In the presence of normal distributions, the one-way ANOVA problem is solved by Snedecor's well known $F$ test. But the assumption of normality cannot be always maintained, because sometimes it is not plausible and at other times it is not verifiable. Assuming the existence, in $H_0$, of a common nondegenerate, continuous, unknown distribution $F$, the problem may be solved by a rank based method such as the Kruskal–Wallis test (Pesarin and Salmaso, 2010). Suppose that $X_1 = (X_{11}, \ldots, X_{1n_1})', \ldots, X_C = (X_{C1}, \ldots, X_{Cn_C})'$ are $C$ samples from independent populations with absolutely continuous CDFs $F_1(x), \ldots, F_C(x)$. Let $R_{ji} = \mathbb{R}(X_{ji}) = \sum_{r=1}^{C} \sum_{u=1}^{n_r} \mathbb{I}(X_{ru} \leq X_{ji})$ denote the rank of $X_{ji}$ in the pooled sample and

$$R_j = \sum_{i=1}^{n_j} R_{ji}$$

$j = 1, \ldots, C$. The Kruskal–Wallis statistic $F_{KW}$ is defined as

$$F_{KW} = \frac{12}{n(n+1)} \sum_{j=1}^{C} \frac{1}{n_j} \left( R_j - n_j \frac{n+1}{2} \right)^2 = \frac{12}{n(n+1)} \sum_{j=1}^{C} \frac{R_j^2}{n_j} - 3(n+1),$$

where $n = \sum_j n_j$ and $(n+1)/2 = \left( \sum_j \sum_i R_{ji}/n \right)$. Note that the Kruskal–Wallis statistic is defined in a similar way to the Snedecor-Fisher statistic by using ranks instead of the original values $X_{ji}$. For large sample sizes the test statistic $F_{KW}$ approximately follows the chi-square distribution with $C - 1$ degrees of freedom under the hypothesis $H_0$ (Kruskal and Wallis, 1952; Bagdonavicius *et al.*, 2011).

Since small values of $F_{KW}$ represent agreement with $H_0$, the hypothesis of null treatment effect should be rejected, at the $\alpha$ level of significance, for large values of the test statistic, that is when $F_{KW} \geq f_{KW,\alpha}$. The critical value $f_{KW,\alpha}$ ensures that the probability of rejecting $H_0$ when it is true (type I error) is equal to $\alpha$. Values of $f_{KW,\alpha}$ are tabulated in the presence of small sample sizes. When the sample sizes are large and then the asymptotic distribution can be used, the critical value is given by the quantile of the chi-square distribution $\chi^2_{C-1,\alpha}$. Of course the usual rejection rule based on the $p$-value can be applied, by computing the $p$-value as probability that the test statistic in $H_0$ assumes greater or equal values than the observed one and comparing it with $\alpha$. In the presence of ties, the average of ranks corresponding to the position values the tied observations cover, should be assigned to the observations in the tied group. This modification may be applied both to the standard procedure and to the large-sample approximation. The exact distribution of the test statistic in the presence of ties has been discussed by Klotz and Teng (1977). In Meyer and Seaman (2013) exact probability distributions for sample sizes up to 35 in each of three groups ($n \leq 105$) and up to 10 in each of four groups ($n \leq 40$) for the test statistic of the Kruskal–Wallis test, are tabulated. The authors compare the exact distributions to the chi-square, gamma, and beta approximations and conclude that in some situations the most common chi-square approximation is not the best choice.

The Kruskal–Wallis test can be considered the multisample extension of the Wilcoxon–Mann–Whitney two-sample test. When the assumption of equal variances is relaxed (multisample Behrens–Fisher problem) the described procedure cannot be applied. A modified test statistic has been proposed by Rust and Fligner (1984) to deal with the multisample Behrens–Fisher problem. Alhakim and Hooper (2008) introduce a large sample nonparametric test for the hypothesis of equal distributions of three or more independent samples, capable of distinguishing between different distributions having the same mean.

Consider for instance the problem related to companies that pack sugar sachets. Each sachet should contain 5 g of sugar. We are interested in comparing the weights of sachets packed by five different companies. We test the hypothesis that the five distributions of sachets' weights are equal against the hypothesis that the central tendencies of the distributions are not equal (i.e., some companies tend to produce sachets with greater or lower weights). We choose a significance level $\alpha = 0.10$. By denoting with $Weight_A$ the random variable which represents the weight of a sachet packed by Company $A$ and using a similar notation for the other companies, the hypotheses of the problem are

$$H_0 : Weight_A \overset{d}{=} Weight_B \overset{d}{=} Weight_C \overset{d}{=} Weight_D \overset{d}{=} Weight_E$$

against the alternative

$$H_1 : H_0 \text{ is not true.}$$

From the production of each company a random sample of 10 sachets has been drawn, and the weights have been recorded (Table 3.1). Formally we have $C = 5$ groups of observations and 10 observations for each group, that is the sample sizes are $n_1 = \cdots = n_5 = 10$.

The $R$ function `kruskal.test(x,g,...)` performs the Kruskal–Wallis test described above. The argument x is a vector or a list of vectors of numeric data and g is a vector or factor object giving the group for the corresponding elements of x.

Table 3.1    Weights (in grams) of sugar sachets packed by five different companies.

| Sachet | Company | | | | |
|--------|------|------|------|------|------|
|        | A    | B    | C    | D    | E    |
| 1      | 5.33 | 5.30 | 4.36 | 3.73 | 4.40 |
| 2      | 4.49 | 5.73 | 4.73 | 3.94 | 5.71 |
| 3      | 5.22 | 4.87 | 6.18 | 5.66 | 5.36 |
| 4      | 4.87 | 5.36 | 5.14 | 4.20 | 5.41 |
| 5      | 4.66 | 4.92 | 5.76 | 4.17 | 4.75 |
| 6      | 5.18 | 6.30 | 3.80 | 4.78 | 4.63 |
| 7      | 5.38 | 4.86 | 4.65 | 3.67 | 4.57 |
| 8      | 5.44 | 3.70 | 4.70 | 5.51 | 5.25 |
| 9      | 4.81 | 5.28 | 5.83 | 4.70 | 5.65 |
| 10     | 5.12 | 5.23 | 5.97 | 4.62 | 5.12 |

It is ignored if x is a list of vectors. In this case the elements of x are vectors of data considered as the samples to be compared, hence they must be numeric data vectors. In this situation one can simply use the command kruskal.test(x) to perform the test. You may also use the command list(x1,...,xn) to create the list of vectors from the vectors x1,...,xn. Otherwise, if vector x is a numeric data vector containing all the sample observations, you need g, vector or factor of the same length as x denoting the group or sample of each element of x. The method automatically adjusts for ties. For the problem of sugar sachets data are imported from the file sugar.csv where the observations are stored in a matrix with $n_j = 10$ rows and $C = 5$ columns as in Table 3.1. The R code to perform the preliminary descriptive analysis is:

```
> sugar=read.csv("sugar.csv",header=TRUE,sep=";")
> attach(sugar)
> weight=c(A,B,C,D,E)
> company=factor(rep(1:5,each=10),labels=c("A","B","C","D","E"))
> boxplot(weight,company,xlab="Company",ylab="Weight")
```

After the data import from the file sugar.csv into the variable sugar, the command attach allows the objects in the database to be accessed by simply giving their names (labels). Hence the instruction weight=c(A,B,C,D,E) allows to create one vector with the $5 \times 10 = 50$ observed data, by stacking up the 5 columns of sugar (here denoted by the column headings A,...,E). With the command company=factor(...) the vector of 50 labels denoting the group for each observation (i.e., the company which packed each sachet) is created. The final command draws the boxplots of data (Figure 3.1). According to the graph it is evident that, from the descriptive point of view, the distributions of weights of sugar sachets seem to be not identical. The sachets of company

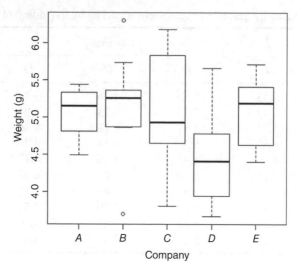

*Figure 3.1    Box-plots of the distributions of weights of sugar sachets packed by five different companies.*

$B$ present two outliers (one upper and one lower), companies $C$ and $D$ pack sachets with higher weight variability and the median weight value for company $D$ seems to be lower and further from the target 5 than the other companies. In any case, it seems that no outstanding differences between the locations of the distributions are present.

To test significant differences in the locations of the distributions, assuming equal variances for the respective populations, the following syntax should be used:

```
> kruskal.test(weight,company)
```

or equivalently

```
> kruskal.test(list(A,B,C,D,E))
######################################################################
#Kruskal-Wallis rank sum test
#
#data:   x and g
#Kruskal-Wallis chi-squared = 6.2692, df = 4,
#                      p-value = 0.1799
######################################################################
```

The kruskal.test function returns the value of the Kruskal–Wallis test statistic, the degrees of freedom of the approximate chi-square distribution of the statistic and the $p$-value of the test. In this case the value of the test statistics is 6.2692 and the corresponding $p$-value is $0.180 > 0.10 = \alpha$. Hence the null hypothesis of no difference in the distributions is not rejected at the significance level 0.10.

As another example of a problem that can be solved with the Kruskal–Wallis test, we consider a pharmacological experiment on the effects of an antipyretic drug on the body temperature of guinea pigs. Four different types of drug (denoted by $T_1$, $T_2$, $T_3$, and $T_4$), with the same active ingredient present in four different concentration levels, were administered to four distinct groups of animals. The goal of the experiment consists in testing whether the effectiveness of the drug differs from group to group, in other words whether the concentration of active ingredient does affect the effectiveness of the drug (significance level $\alpha = 0.01$).

Formally, by denoting with $Temp_j$ the random variable which represents the temperature variation of a guinea pig after the administration of drug $T_j$ ($j = 1, \dots, 4$), the hypotheses of the problem are

$$H_0 : Temp_1 \overset{d}{=} Temp_2 \overset{d}{=} Temp_3 \overset{d}{=} Temp_4$$

against the alternative

$$H_1 : H_0 \text{ is not true.}$$

Four independent samples of 15 guinea pigs were involved in the experiment and each of the four treatments was administered to a distinct group. Hence we have $C = 4$ and $n_1 = \dots = n_4 = 15$. The temperature change (in °C) after the administration of the drug observed on each animal is reported in Table 3.2.

Table 3.2    Body temperature change (in °C) in guinea pigs after the administration of an antipyretic drug with four different concentration levels of active ingredient.

| Animal | Drug | | | |
|---|---|---|---|---|
| | $T_1$ | $T_2$ | $T_3$ | $T_4$ |
| 1 | 1.4 | −1.7 | 0.2 | −3.9 |
| 2 | −1.2 | −3.3 | 1.1 | −2.6 |
| 3 | −0.8 | −2.4 | 1.8 | −3.0 |
| 4 | 0.2 | −1.9 | 1.9 | −4.4 |
| 5 | −0.1 | −1.9 | −0.6 | −2.7 |
| 6 | −0.3 | −4.2 | 0.8 | −2.3 |
| 7 | 1.5 | −1.8 | −1.1 | −1.2 |
| 8 | 0.4 | −2.7 | −0.8 | −2.3 |
| 9 | −0.2 | −2.6 | −1.1 | −3.3 |
| 10 | 2.9 | −2.1 | −1.8 | −5.6 |
| 11 | 1.1 | −3.1 | 0.3 | −3.0 |
| 12 | 0.2 | −2.3 | 1.7 | −3.6 |
| 13 | −1.3 | −1.6 | −1.1 | −4.2 |
| 14 | −0.8 | −2.1 | 1.5 | −4.2 |
| 15 | 0.5 | −3.7 | −0.3 | −1.7 |

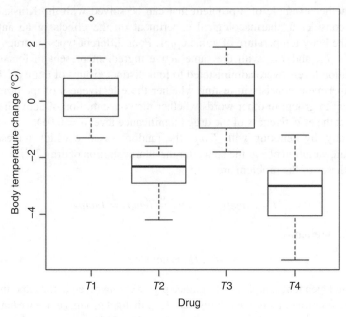

*Figure 3.2    Box-plots of body temperature change in guinea pigs after the administration of an antipyretic drug with four different concentration levels of active ingredient.*

The data of this problem are stored in the file `drugs.csv`. To represent the data with the box-plots we use the $R$ commands shown before:

```
> drugs=read.csv("drugs.csv",header=TRUE,sep=";")
> attach(drugs)
> temp=c(T1,T2,T3,T4)
> drug=factor(rep(1:4,each=15),labels=c("T1","T2","T3","T4"))
> boxplot(temp,drug,xlab="Drug",ylab="Body temperature")
```

According to the box-plots, from the descriptive point of view, the distributions seem to be different, mainly because the location of the distributions of temperature changes for drugs $T_1$ and $T_3$ are around zero while that of the other distributions are negative, that is only two treatments seem to have non-null effects (Figure 3.2).

The application of the usual commands allows the Kruskall–Wallis test to be performed:

```
> kruskal.test(temp,drug)
```

or alternatively

```
> kruskal.test(list(T1,T2,T3,T4))
######################################################################
#Kruskal—Wallis rank sum test
#
#data:  x and g
#Kruskal—Wallis chi-squared = 43.9763, df = 3,
#                 p-value = 1.527e-09
######################################################################
```

In this problem the $p$-value is almost 0. Thus the null hypothesis of equal distributions between the four treatments is rejected in favor of the alternative hypothesis of different effects of the four treatments.

## 3.2.2    Permutation ANOVA in the presence of one factor

An alternative solution for the problem described in the previous subsection, that is the testing problem for a $C$-sample one-way ANOVA, may be found within the permutation procedures. Note that because of conditioning to a set of sufficient statistics, this procedure allows for relaxation of continuity for $F$ and for the condition of finite variances for responses. It only requires the existence of location coefficients and of proper sampling estimators for them (Pesarin and Salmaso, 2010). Suppose that $X_1 = (X_{11}, \ldots, X_{1n_1})', \ldots, X_C = (X_{C1}, \ldots, X_{Cn_C})'$ are $C$ samples from independent populations with nondegenerate distribution functions $F_1(x), \ldots, F_C(x)$. Under the homoscedastic fixed effects model the hypotheses are

$$H_0 : F_1(x) = \ldots = F_C(x)$$

against

$$H_1 : \text{at least one } F_j(x) \text{ is different from the others,}$$

with $j = 1, \ldots, C$. Equivalently we can write $H_0 : \theta_1 = \cdots = \theta_C$ and $H_1 :$ some $\theta_j$ are not equal, where $\theta_j, j = 1, \ldots, C$ are the effects of the $C$ treatments. The permutation test is more flexible than the Kruskall–Wallis test because it can be applied not only to continuous variables but also to discrete ones. Furthermore for the application of the permutation test the existence of mean values is not necessary. The only necessary condition is the existence of location parameters and of sample estimators for them. Without loss of generality let us assume that the sample means are proper indicators of treatment effects. A suitable test statistic based on deviance among sample means is

$$T_{Dev}^* = \sum_{j=1}^{C} \left( \overline{X}_j^* - \overline{X}_. \right)^2 \cdot n_j,$$

where $\overline{X}_j^* = \frac{1}{n_j} \sum_i X_{ji}^*$ are permutation sample means related to the permuted dataset, and $\overline{X}_. = \frac{1}{n} \sum_j n_j \cdot \overline{X}_j$. Note that $\overline{X}_.$ is a permutationally invariant quantity, being based on the sum of all observed data. Hence, statistic $T^*_{Dev}$ is permutationally equivalent to $T^* = \sum_{j=1}^{C} (\overline{X}^*_j)^2 \cdot n_j$. Under the null hypothesis data are exchangeable among groups, hence the permutation distribution of the test statistic can be obtained or estimated by permuting $B$ times the observations in the dataset and computing the $B$ corresponding values of the statistic. The null hypothesis must be rejected for high values of the test statistic and consequently the $p$-value is the probability that the test statistic takes greater values than the observed one according to the permutation distribution.

Consider again the problem of the companies that pack sugar sachets where we are interested in testing whether the weights of the sachets packed by five different companies are equal or not (Table 3.1 and Figure 3.1). The $R$ code for importing the data from the file sugar.csv, for defining the vector of observations of the response variable weight and the vector of labels denoting the sample company are the same considered for the Kruskal–Wallis test:

```
> sugar=read.csv("sugar.csv",header=TRUE,sep=";")
> attach(sugar)
> company=factor(rep(1:5,each=10),labels=c("A","B","C","D","E"))
> weight=c(A,B,C,D,E)
```

To apply the permutation solution we use the script for the computation of the significance level function (t2p.r) and for the estimation of the permutation distribution of the test statistic (oneWayPerm.r) with the source command:

```
> source("t2p.r")
> source("oneWayPerm.r")
```

The test can be performed as follows:

```
> one.way.perm(weight,company,B=10000)
```

The output is:

```
######################################################################
# Permutation solution
# $n
# Y
# A  B   C   D   E
# 10 10 10 10 10
#
# $C
# [1] 5
#
# $pvalue
# [1] 0.1174883
######################################################################
```

The $R$ function one.way.perm(X,Y,B) needs as input the vector of data X, the vector of the group labels Y and the number of permutations B. The function returns the following list of data:

- $n: sample sizes for each group,

- $C: number of samples,

- $pvalue: permutation $p$-value.

For the considered problem, the resulting $p$-value 0.117 is greater than $\alpha$, thus the null hypothesis of no difference between the five distributions should not be rejected.

The problem concerning the pharmacological experiment related to an antipyretic drug administered to four groups of guinea pigs in different concentration levels of active ingredient can also be solved with a permutation procedure. The data (differences in body temperature before and after the administration of the drug) and their representation with box-plots are shown in Table 3.2 and Figure 3.2 respectively. The $R$ code for importing data from the file drugs.csv and to perform the analysis is:

```
> drugs=read.csv("drugs.csv",header=TRUE,sep=";")
> attach(drugs)
> drug=factor(rep(1:4,each=15),labels=c("T1","T2","T3","T4"))
> temp=c(T1,T2,T3,T4)
> source("t2p.r")
> source("oneWayPerm.r")
> one.way.perm(temp,drug).
```

The corresponding output is:

```
####################################################################
# Permutation solution
# $n
# Y
#  T1  T2  T3  T4
#  15  15  15  15
#
# $C
# [1] 4
#
# $pvalue
# [1] 9.999e-05
####################################################################
```

In this case, the $p$-value is almost 0 and leads to the rejection of the null hypothesis of no difference in distribution between the $C = 4$ treatments in favor of the alternative that at least one treatment produces an effect which differs from the others.

Another goal of this problem could be to test for increasing relation between concentration levels of active ingredient and effects. In this case a stochastic ordering analysis is useful and possible using the combination based permutation tests.

### 3.2.3    The Mack–Wolfe test for umbrella alternatives

In a one-way ANOVA design, the response may stochastically increase with the treatment level up to a point and then decrease. This 'up-then-down behavior' is known in the literature as *umbrella ordering* (Mack and Wolfe, 1981). In previous sections we have seen procedures for testing $H_0 : \theta_1 = \theta_2 = \cdots = \theta_C$ against the general alternative that at least one equality is not true, that is not all the effects are equal. Now let us take into account procedures for testing $H_0$ against the class of umbrella alternatives. Let $p \in \{1, \ldots, C\}$ be a fixed treatment label. The alternative hypothesis of the problem is

$$H_1 : \theta_1 \leq \cdots \leq \theta_{p-1} \leq \theta_p \geq \theta_{p+1} \geq \cdots \geq \theta_C,$$

where at least one strict inequality is true. This alternative is said to have a peak at population $p$. This situation is common in many real problems, such as clinical studies, where it is of interest to study the reaction to increasing drug dosage levels where an initial increasing effect culminating with a peak point (corresponding to the optimal dosage) and a subsequent decreasing effect may occur. Other typical *umbrella orderings* concern the effect of age on some physiological parameters that measure the physical efficiency, such as the *resilience after physical exertion, muscle strength, cardiac efficiency*, etc. or the effect of increasing percentages of a given additive on the physical–mechanical properties of certain materials. A solution to these problems was provided by Mack and Wolfe (1981). The application of the method requires a preliminary distinction between two possible procedures depending on whether the peak is known or unknown.

Let us consider the former procedure. This method is preferred to the Kruskal–Wallis test when in the alternative hypothesis we do not just indicate that not all the effects are equal but that the treatments can be labeled in such a way that specific deviations from $H_0$ are expected by the treatment effects. According to these deviations, the effects of the first $p$ populations are non decreasing with respect to the treatment levels and the other $C - p$ are non increasing, where $p$ is known (Hollander and Wolfe, 1999). Hence we must label the treatments to obtain the prescribed ordered relationship corresponding to the umbrella configuration in $H_1$. Then we must compute the $p(p - 1)/2$ Mann–Whitney counts given by

$$U_{uv} = \sum_{i_1=1}^{n_u} \sum_{i_2=1}^{n_v} \phi\left(X_{ui_1}, X_{vi_2}\right),$$

where $\phi(a, b) = 1$ if $a < b$, 0 otherwise, for $1 \leq u < v \leq p$ and the $(C - p + 1)$ $(C - p)/2$ reverse Mann–Whitney counts $U_{vu}$ for $p \leq u < v \leq C$. The *peak-known*

Mack–Wolfe umbrella statistic is

$$MK_p = \sum_{u=1}^{v-1} \sum_{v=2}^{p} U_{uv} + \sum_{u=p}^{v-1} \sum_{v=p+1}^{C} U_{vu}.$$

To test $H_0$ versus the *peak-known* (at $p \in \{1, \ldots, C\}$) umbrella alternative $H_1$ at $\alpha$ significance level, let $a_{p,\alpha}$ be the critical value such that the type I error probability is equal to $\alpha$ (for selected $p$ and $C$ values), then the null hypothesis should be rejected when $MK_p \geq a_{p,\alpha}$. Values of $a_{p,\alpha}$ are tabulated (see Appendix A).

When $\min(n_1, \ldots, n_C)$ tends to infinity we can consider an approximation for the null distribution of $MK_p$. Consider the standardized version of $MK_p$:

$$\widetilde{MK}_p = \frac{MK_p - \mathbb{E}_0(MK_p)}{\sqrt{\mathbb{V}_0(MK_p)}},$$

where

$$\mathbb{E}_0(MK_p) = N_1^2 + N_2^2 - \sum_{j=1}^{C} n_j^2 - n_p^2$$

and

$$\mathbb{V}_0(MK_p) = \frac{1}{72} \left[ 2 \left( N_1^3 + N_2^3 \right) + 3 \left( N_1^2 + N_2^2 \right) - \sum_{j=1}^{C} n_j^2 (2n_j + 3) \right.$$
$$\left. - n_p^2 (2n_p + 3) + 12 n_p N_1 N_2 - 12 n_p^2 n \right],$$

where $\mathbb{E}_0$ and $\mathbb{V}_0$ denote the mean and the variance of the test statistic under $H_0$, respectively, with $N_1 = \sum_{j=1}^{p} n_j$ and $N_2 = \sum_{j=p}^{C} n_j$. Note that $n = N_1 + N_2 - n_p$ since the $n_p$ observations in the peak treatment $p$ are counted in both $N_1$ and $N_2$. When $H_0$ is true, $\widetilde{MK}_p$ is asymptotically $\mathcal{N}(0, 1)$ distributed as $\min(n_1, \ldots, n_C)$ tends to $\infty$, then the null hypothesis must be rejected when $\widetilde{MK}_p \geq z_\alpha$, where $z_\alpha$ is the $\alpha$-quantile of the standard normal distribution. If there are ties among the $N_1$ observations in treatments $1, \ldots, p$ or in the $N_2$ ones in treatments $p, \ldots, C$, we proceed by replacing the function $\phi(a, b)$ in the computations of the appropriate Mann–Whitney counts $U_{uv}$ or reverse Mann–Whitney counts $U_{vu}$ with $\phi^|(a, b) = 1, \frac{1}{2}, 0$ if $a < b, a = b$, or $a > b$, respectively. For the modified $MK_p$ in the case of ties, we can refer to the same critical value $a_{p,\alpha}$. It is worth noting that the significance level of the test in the presence of tied observations is only approximately, and not exactly, equal to $\alpha$. Note that, even when using the approximation for large samples, we have to take into account the possible presence of ties. In particular, under $H_0$ in the case of ties, the true null variance is smaller than the numerical value given above. However, the appropriate expression for the exact variance of $MK_p$ in the case of ties is not available. Therefore, in the case of tied observations and large sample sizes, it is recommended to compute $MK_p$ using the modified Mann–Whitney counts and then $\widetilde{MK}_p$.

Let consider now the procedure for testing $H_0 : \theta_1 = \theta_2 = \cdots = \theta_C$ against the general *peak-unknown* umbrella alternatives

$$H_1 : \theta_1 \leq \cdots \leq \theta_{p-1} \leq \theta_p \geq \theta_{p+1} \geq \cdots \geq \theta_C$$

with at least one strict inequality, for some $p \in \{1, 2, \dots, C\}$. In this case we first use the sample data to estimate the pick treatment, that is we etimate $p$ from the observed data. To do this we have to compute $C$ combined samples Mann–Whitney statistics

$$U_{.v} = \sum_{u \neq v} U_{uv}$$

for $v = 1, \dots, C$. Note that, $U_{.v}$ is itself simply a single Mann–Whitney statistic computed between the $v$th sample and the remaning $(C - 1)$ samples combined. Then, standardization of the statistics should be applied as follows:

$$\widetilde{U}_{.v} = \frac{U_{.v} - \mathbb{E}_0(U_{.v})}{\sqrt{\mathbb{V}_0(U_{.v})}}$$

with $v = 1, \dots, C$,

$$\mathbb{E}_0(U_{.v}) = \frac{n_v(n - n_v)}{2}$$

and

$$\mathbb{V}_0(U_{.v}) = \frac{n_v(n - n_v)(n + 1)}{12}.$$

Let $A$ be the subset of $\{1, 2, \dots, C\}$ with the $r$ treatments tied for having the maximum value of $\widetilde{U}_{.v}$. Then, the *peak-unknown* Mack–Wolfe statistic is given by

$$\widetilde{MU}_{\widehat{p}} = \frac{1}{r} \sum_{j \in A} \frac{MK_j - \mathbb{E}_0(MK_j)}{\sqrt{\mathbb{V}_0(MK_j)}},$$

where $MK_j$ is the *peak-known* statistic with the peak at the $j$th treatment group. To test $H_0$ against the *peak-unknown* umbrella alternative at $\alpha$ significance level we reject the null hypothesis if $\widetilde{MU}_{\widehat{p}} \geq \widehat{a}_{\widehat{p},\alpha}$. The critical value $\widehat{a}_{\widehat{p},\alpha}$ is chosen to make the type I error probability equal to $\alpha$. The values of $\widehat{a}_{\widehat{p},\alpha}$ are tabulated (see Appendix B). In the presence of ties among the observations, as in the case of *peak-known* alternative, $\phi(a, b)$ in the computation of the associated Mann–Whitney counts $U_{uv}$ or reverse Mann–Whitney counts $U_{vu}$ should be replaced by $\phi^|(a, b) = 1, \frac{1}{2}, 0$ if $a < b$, $a = b$, or $a > b$, respectively. After computing $U_{.v}$ and $\widetilde{MU}_{\widehat{p}}$ with this correction for ties, we can refer to the distribution of $\widetilde{MU}_{\widehat{p}}$, noting that this adjusted test has a significance level approximately equal to $\alpha$.

Let us consider a medical experiment. A cure based on a treatment for people with difficulties using their jaw was applied to patients classified into four age groups

Table 3.3   Maximum assisted mouth opening (in millimeters) after the cure for people with difficulties using their jaw by age group.

| 'Age group (yr)' | | | |
|---|---|---|---|
| ≤30 | (30 − 45] | (45 − 64] | ≥65 |
| 57.71 | 48.54 | 61.00 | 36.61 |
| 39.14 | 48.74 | 48.34 | 41.61 |
| 48.40 | 42.28 | 58.99 | 46.82 |
| 41.85 | 47.29 | 68.28 | 34.95 |
| 24.72 | 60.09 | 44.19 | 42.08 |
| 40.71 | 33.35 | 46.85 | 53.92 |
| 44.42 | 33.19 | 52.48 | 46.04 |
| 52.18 | 35.44 | 50.33 | 51.70 |
| 43.28 | 41.05 | 36.44 | 46.70 |
| 52.64 | 48.52 | 43.22 | 36.76 |

(ordered categories), who presented the same level of maximum assisted mouth opening (MAMO). The goal of the study is to test whether after the cure the MAMO tends to increase with age up to a peak and then to decrease for higher ages. In other words, we wish to test if the effect of the cure presents an umbrella ordering with respect to age. By denoting with $MAMO_j$ the random variable representing the MAMO for the $j$th age group the problem consists of testing

$$H_0 : MAMO_1 \overset{d}{=} \dots \overset{d}{=} MAMO_4$$

against

$$H_1 : MAMO_1 \overset{d}{\leq} \dots \overset{d}{\leq} MAMO_p \overset{d}{\geq} \dots \overset{d}{\geq} MAMO_4$$

with $p$ unknown. The experimental data are displayed in Table 3.3.

Descriptive analysis based on the box-plots (Figure 3.3) indicates that differences in the effects of the cure between the groups are not evident. A slight increase of the effect can be seen up to the third group (from 45- to 64-years-old patients) that could represent the peak point. The $R$ code to import data from the file MAMO.csv and represent the distributions with box-plots is:

```
> data=read.csv("MAMO.csv",header=TRUE,sep=";")
> x=c(data[,1],data[,2],data[,3],data[,4])
> y=rep(1:ncol(data),each=nrow(data))
> boxplot(x,y,xlab="Age group",ylab="MAMO-Maximum Assisted Mouth
  Opening")
```

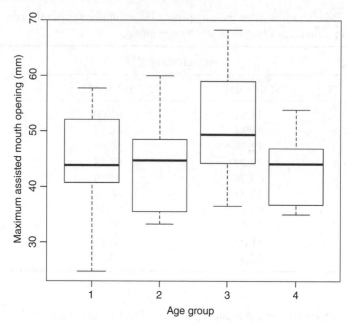

*Figure 3.3    Box-plots of maximum assisted mouth opening after the cure for people with difficulties using their jaw by age group.*

The $R$ function MW(x,y), available in the file "MW.r", computes the *peak-unknown* Mack–Wolfe statistic. This function requires as inputs the vector of observations x and a further vector of labels denoting the groups y. The $R$ code and the final output are:

```
> source("MW.r")
> MW(x,y)
##################################################################
#Mack and Wolfe solution
#
# $T
# 1.703183
#
# $A
# -0.09654216
#   0.02886751
#   1.70318329
#   0.09654216
#
# $Z
#  -0.2498780
#  -0.7183993
#   1.9365546
```

```
#   -0.9682773
#
# $peak
#   3
#####################################################################
```

The function MW returns a list of results. In particular A contains the $C = 4$ *peak-known* statistics $MK_j$ ($j = 1, \ldots, C$) computed for each group, Z contains the $C = 4$ standardized combined samples Mann–Whitney statistics $\widetilde{U}_{.v}$ ($v = 1, \ldots, C$). Thus, the value of the *peak-unknown* Mack–Wolfe statistic T corresponds to the value of A computed for the group with maximum value of Z. Therefore the corresponding group (in this case the third) is the estimated peak group (peak). To solve the testing problem, the observed value of the statistic must be compared with the critical value $\widetilde{a}_{3,0.05} = 2.172$. Since the critical value is greater than $T = 1.703$, the null hypothesis of equal effects of the treatment for each group should not be rejected.

Consider now an engineering study on the mechanical characteristics of fiber-reinforced concrete. The study consists of testing whether the tensile strength is an increasing function of the percentage of a specific additive up to a certain (unknown) percentage level (peak point), and a decreasing function of the percentage after this point. Let us assume $C = 5$ samples with 6 observations about to break and that each sample corresponds to a specific additive percentage. Table 3.4 shows the observed experimental data.

The *R* code for the analysis is:

```
> source("MW.r")
> data=read.csv("concrete.csv",header=TRUE,sep=";")
> x=c(data[,1],data[,2],data[,3],data[,4],data[,5])
> y=rep(1:ncol(data),each=nrow(data))
> boxplot(x,y,xlab="Additive percentage",ylab="Tensile Strength")
> MW(x,y)
```

Table 3.4   Tensile strength (in N/m$^2$) of fiber-reinforced concretes by percentage of a specific additive.

| 'Percentage of additive' | | | | |
|---|---|---|---|---|
| 16 | 20 | 24 | 28 | 32 |
| 5.00 | 3.77 | 4.83 | 8.94 | 4.78 |
| 3.12 | 6.22 | 6.43 | 6.57 | 3.04 |
| 2.82 | 6.84 | 7.29 | 6.63 | 4.48 |
| 4.19 | 5.16 | 7.77 | 6.75 | 5.87 |
| 5.26 | 5.33 | 6.62 | 8.10 | 5.26 |
| 2.76 | 7.03 | 5.12 | 8.02 | 5.83 |

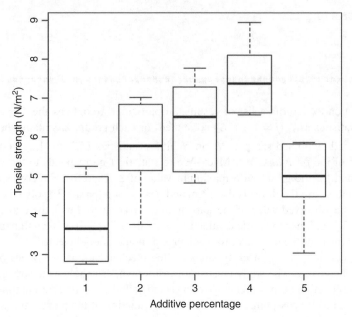

*Figure 3.4   Box-plots of tensile strength of fiber-reinforced concretes by percentage of a specific additive.*

The box-plots in Figure 3.4 show that the distributions of the tensile strength seem to increase up to the fourth sample (additive percentage at level 4). The final output of the inferential analysis is:

```
####################################################################
# Mack and Wolfe solution
#
# $T
# 4.015209
#
# $A
# -1.643168
#  1.091089
#  2.451304
#  4.015209
#  1.679683
#
# $Z
#  -2.9034647
#   0.3110855
#   1.0368517
```

```
#     3.0590075
#    -1.4517324
#
# $peak
#    4
################################################################################
```

In this example, the critical value at level $\alpha = 0.05$ is $\widetilde{a}_{4,0.05} = 2.227$. Since it is smaller than the observed value of the test statistic $T = 4.015$ the null hypothesis must be rejected at the significance level $\alpha = 0.05$. Therefore we conclude that there is empirical evidence supporting the alternative hypothesis that the tensile strength has an umbrella behavior with respect to the percentage of additive and the peak point corresponds to the fourth group that is to 28% of additive (where $\widetilde{U}_{.y}$ reaches its maximum 3.059).

### 3.2.4    Permutation test for umbrella alternatives

The method presented in the previous subsection for multisample tests with umbrella alternatives, suggested by Mack and Wolfe (1981), is a nonparametric rank solution. In this procedure the test statistic is a weighted linear combination of standardized Mann–Whitney statistics, and the authors compute the null distribution in a wide variety of situations. Basso and Salmaso (2011) proposed an alternative solution based on a permutation approach conditional on the observed data considered as a set of sufficient statistics in $H_0$. The null hypothesis can be written as

$$H_0 : F_1(x) = F_2(x) = \cdots = F_C(x) \; \forall x \in \mathcal{R}$$

against the umbrella alternative

$$H_1 : F_1(x) \geq \cdots \geq F_{p-1}(x) \geq F_p(x) \leq F_{p+1}(x) \leq \cdots \leq F_C(x)$$

for some $p \in \{1, 2, \ldots, C\}$ and with at least one strict inequality, where $F_j(x)$ is the CDF of the random variable representing the response for the $j$th population. When the peak group, corresponding to population $p$, is known, the alternative hypothesis can be represented as the intersection of two simple stochastic ordering alternatives (one increasing and one decreasing). The solution can be found by combining two partial tests for simple stochastic ordering alternatives, that is

$$H_1 : H_{1p}^{\nearrow} \cap H_{1p}^{\searrow},$$

where $H_{1p}^{\nearrow} = F_1(x) \geq \cdots \geq F_p(x)$ and $H_{1p}^{\searrow} = F_p(x) \leq \cdots \leq F_C(x)$ (Pesarin and Salmaso, 2010). The alternative hypothesis $H_1$ is true when both $H_{1p}^{\nearrow}$ and $H_{1p}^{\searrow}$ are true. Let us consider the null hypothesis $H_0 : F_1(x) = \cdots = F_p(x)$ against the first partial alternative $H_{1p}^{\nearrow}$. Note that if $p$ is equal to 2 the problem reduces to a simple one-tailed, two-sample location problem. If $p > 2$ let us consider all the possible

$p - 1$ subdivisions of the dataset into two disjoint pseudo-groups, where the first group, say $Y_{(1)s}$, is obtained by pooling together data of the first $s$ samples (ordered with respect to the treatment levels), and the second, $Y_{(2)s}$, by pooling together the remaining observations, for $s = 1, \dots, p - 1$. Assume that, for each $s = 1, \dots, p - 1$, data of the $j$th pseudo-group $Y_{(j)s}$ are realizations of a random variable with CDF $F_{(j)s}$, with $j = 1, 2$. Hence the problem related to the partial test on increasing stochastic ordering can be broken down into $p - 1$ sub-problems and the hypotheses can be formulated as

$$H_{0p}^{\nearrow} : \cap_{s=1}^{p-1} [F_{(1)s}(x) = F_{(2)s}(x)] \; \forall x \in \mathcal{R} \equiv \cap_{s=1}^{p-1} H_{0s}$$

against

$$H_{1p}^{\nearrow} : \cup_{s=1}^{p-1} [F_{(1)s}(x) \geq F_{(2)s}(x)] \equiv \cup_{s=1}^{p-1} H_{1s},$$

where the null hypothesis implies that all the null sub-hypotheses $F_{(1)s}(x) = F_{(2)s}(x)$ are true and the alternative hypothesis states that at least one alternative sub-hypothesis $F_{(1)s}(x) \geq F_{(2)s}(x)$ is true and for some $x$ the strict inequality is true. The $s$th sub-problem corresponds to a two-sample comparison for restricted alternatives which has an exact and unbiased permutation solution, using the test statistics

$$T_{s\nearrow}^* = \frac{\overline{Y}_{(2)s}^* - \overline{Y}_{(1)s}^*}{\sqrt{\hat{\sigma}_s^2 \left( \dfrac{1}{n_{(1)s}} + \dfrac{1}{n_{(2)s}} \right)}}$$

$s = 1, \dots, p - 1$ where $\overline{Y}_{(2)s}^*$ and $\overline{Y}_{(1)s}^*$ are the sample means of the second and first pseudo-groups, respectively, in a permuted dataset, $\hat{\sigma}_s^2$ is the estimated variance of the pooled dataset, and $n_{(1)s}$ and $n_{(2)s}$ are the sizes of $Y_{(1)s}$ and $Y_{(2)s}$, respectively, that is the sample sizes of the $s$th sub-problem. Large values of the test statistic $T_{s\nearrow}^*$ lead to the rejection of $H_{0s}$ in favor of the alternative $H_{1s}$. A combination of the significance level functions of the partial tests $T_{s\nearrow}^*$ ($s = 1, \dots, p$), according to the nonparametric combination (NPC) methodology, provides a test statistic for testing $H_{0p}^{\nearrow}$ against $H_{1p}^{\nearrow}$. A similar procedure can be applied to the test $H_{0p}^{\searrow} = F_p(x) = \cdots = F_C(x)$ against $H_{1p}^{\searrow} = F_p(x) \leq \cdots \leq F_C(x)$ and a further combination of the two tests provides a final test statistic and a final $p$-value for the overall testing problem for umbrella alternatives. The NPC methodology again provides a solution to the general testing problem for umbrella alternatives.

When the peak group is unknown (a very common situation in real problems) a solution consists of repeating the procedure for known peak $C$ times: for $p = 1$, $p = 2, \dots, p = C$. Hence, for $j = 1, \dots, C$ let

$$\psi_{j\nearrow}^* = \sum_{s=1}^{j-1} T_{s\nearrow}^* \quad \text{and} \quad \psi_{j\searrow}^* = \sum_{s=j}^{C-1} T_{s\searrow}^*$$

be two statistics to test $H_{0j}^{\nearrow}$ against $H_{1j}^{\nearrow}$ and $H_{0j}^{\searrow}$ against $H_{1j}^{\searrow}$, respectively, obtained with the direct combination (sum of the test statistics) of the partial tests $T_{s\nearrow}^*$ and $T_{s\searrow}^*$. Note that when $j = 1$ we actually test for decreasing stochastic ordering, whereas when $j = C$ we test for increasing stochastic ordering. A solution for the testing problem of umbrella alternatives with unknown peak consists of combining the two partial tests, for example applying the Fisher's combining function on the permutation significance level functions $p_{j\nearrow}''^*$ and $p_{j\searrow}''^*$ of the two partial tests:

$$\Psi_j^* = -2 \cdot \log \left( p_{j\nearrow}''^* \cdot p_{j\searrow}''^* \right).$$

When the $p$-value of this combined test $\lambda_j$ is less than $\alpha$ then there is empirical evidence of umbrella ordering with peak group $j$. In order to evaluate whether there is a significant presence of any umbrella alternative, the $C$ tests for the umbrella ordering with known peak, should be further combined through the NPC methodology. By denoting with $\lambda_j^*$ the significance level function of the $j$th test, a suitable combination can be provided by the application of the Tippett rule:

$$\Psi''^* = \max \left( 1 - \lambda_1^*, \ldots, 1 - \lambda_C^* \right).$$

Large values of $\Psi''$ lead to the rejection of the null hypothesis of equality in distribution of the $C$ populations in favor of the alternative of umbrella ordering. Hence the permutation $p$-value is $\lambda'' = \sum_{b=1}^{B} \mathbb{I}(\Psi_{(b)}''^* \geq \Psi_{obs}'')/B$, where $B$ is the number of permutations, $\Psi_{(b)}''^*$ is the value of the test statistic corresponding to the $b$th permutation, $\Psi_{obs}''$ is the observed value of the test statistic and $\mathbb{I}(\cdot)$ denotes the indicator function. When $\lambda'' < \alpha$ the null hypothesis should be rejected in favor of the umbrella alternative. The peak of the umbrella corresponds to the group with minimum $p$-value.

Consider again the medical experiment where a cure based on a treatment for people with jaw dysfunctions was given to patients classified into four age groups (ordered categories), who presented the same level of MAMO before the treatment (Table 3.3 and Figure 3.3). To test the null hypothesis of equality in distribution of MAMO for different age groups after the treatment against the umbrella alternative (with respect to age) with *unknown peak*, let us apply the permutation procedure using the $R$ function umbrella(x,y,B). This function requires as input three arguments: the vector x of observed data, the vector y of the group labels denoting the sample of each observation, and the number of permutations B. The $R$ syntax for the analysis of data included in the file MAMO.csv and the final output are:

```
> source("umbrella.r")
> source("t2p.r")
> source("combine.r")
> data=read.csv("MAMO.csv",header=TRUE,sep=";")
> x=c(data[,1],data[,2],data[,3],data[,4])
> y=rep(1:ncol(data),each=nrow(data))
> umbrella(x,y,B=10000)
```

```
###########################################################################
# $Global.p.value
# [1] 0.099
#
# $Partial.p.values
# [1] 0.61023898 0.57344266 0.02629737 0.38986101
#
# $Max
# [1] 3
###########################################################################
```

Note that the function umbrella requires the $R$ code included in "t2p.r" (for the computation of the significance level function), the script in "combine.r" (for the NPC of tests) and the file "umbrella.r" for the application of the testing procedure. The function umbrella returns a list of objects: the global $p$-value of the test (Global.p.value), the partial $p$-values related to the sub-tests for *known peak* (Partial.p.values) and the label of the group identified as peak. Note that it corresponds to the group whose partial test presents the lower $p$-value. For this problem the global $p$-value 0.099 is greater than $\alpha = 0.05$, thus the null hypothesis of equality in distribution must not be rejected.

Consider now the engineering study on the mechanical characteristics of the fiber-reinforced concrete where the interest of the study consists of testing whether the tensile strength is an increasing function of the percentage of a specific additive up to a certain (unknown) percentage level (peak point), and a decreasing function for higher percentage levels (Table 3.4 and Figure 3.4). The analysis is performed similarly to the previous example, by importing data from the file "concrete.csv":

```
> source("umbrella.r")
> source("t2p.r")
> source("combine.r")
> data=read.csv("concrete.csv",header=TRUE,sep=";")
> x=c(data[,1],data[,2],data[,3],data[,4],data[,5])
> y=rep(1:ncol(data),each=nrow(data))
> umbrella(x,y,B=10000)
###########################################################################
# $Global.p.value
# [1] 0
#
# $Partial.p.values
# [1] 0.95630437 0.05879412 0.00699930 0.00009999 0.04379562
#
# $Max
# [1] 4
###########################################################################
```

The global $p$-value here is almost 0, hence the null hypothesis at the significance level $\alpha = 0.05$ must be rejected in favor of the umbrella alternative. The minimum partial $p$-value corresponds to group 4, thus there is evidence that the tensile strength is an increasing function of additive percentage up to 28% (fourth percentage level) and a decreasing function of it for higher percentages.

## 3.3    Two-way ANOVA layout

In this section we take into account experimental designs involving two factors at two or more levels. Usually in these problems the primary interest is in the relative location effects of the different levels of one factor, called the treatment factor, within the various blocks of observations defined according to the levels of the second factor, called the blocking factor. This situation is typical of experimental designs where subjects are first divided into homogeneous subgroups (blocks) and then randomly assigned to the various treatment levels. Such a design is called *randomized block design* (Hollander and Wolfe, 1999). In the null hypothesis there is no difference between the location effects of the treatments. In the alternative hypothesis there are significant differences between the treatment effects. By assuming that $K$ blocks and $C$ treatments are taken into account and that, for each block-treatment combination, $n_{kj}$ replications are considered with $k = 1, \ldots, K$ and $j = 1, \ldots, C$, the data consist of the set $\{X_{kji}; \ i = 1, \ldots, n_{kj}; \ k = 1, \ldots, K; \ j = 1, \ldots, C\}$, with $n = \sum_{k,j} n_{kj}$.

Let us assume that for a given block-treatment combination $(k, j)$ $n_{kj}$ observations are randomly sampled from a population with CDF $F_{kj}$. The $K \cdot C$ distributions possibly differ only for the location. Let $X_{kji}$ be a realization of the random variable $Z_{kji}$ with CDF $F_{kj}$, the classical two-way ANOVA layout, according to the additive model representation, is

$$Z_{kji} = \theta + \beta_k + \mu_j + \epsilon_{kji},$$

with $i = 1, \ldots, n_{kj}$, $k = 1, \ldots, K$ and $j = 1, \ldots, C$, where $\theta$ is a population constant (e.g., unknown mean or median), $\mu_j$ is the unknown effect of the $j$th treatment and $\epsilon_{kji}$ are exchangeable random errors with zero mean or median. Parameters $\beta_k$ and $\mu_j$ represent the $k$th block effect and $j$th treatment effect, respectively.

### 3.3.1    The Friedman rank test for unreplicated block design

Sometimes we are in the presence of randomized complete block design, where replications are $n_{kj} = 1$ for every $k$ and every $j$ and the goal is to test $H_0 : [\mu_1 = \cdots = \mu_C]$ against $H_1 : [\mu_1, \ldots, \mu_C$ not all equal]. When the assumption of normality of the classical parametric solution for ANOVA problems is not plausible or when variances of different populations may not be equal, the Friedman test provides a nonparametric solution to this problem based on ranks (Kvam and Vidakovic, 2007). The statistic proposed by Friedman is based on the within block ranks (see Friedman, 1937). No block effect can be tested by this rank-based approach (Pesarin, 2001).

Let $X_{kj}$ be the (only) observation related to the $k$th block and $j$th treatment and $R_{kj}$ be the rank of $X_{kj}$ in the set of observations of the $k$th block $\{X_{k1}, \ldots, X_{kC}\}$: formally $R_{kj} = \mathbb{R}_k(X_{kj})$. Let $R_{\bullet j}$ denote the mean rank for the $j$th treatment, that is $R_j = \sum_k R_{kj}$ and $R_{\bullet j} = R_j/K$. The Friedman test statistic is

$$FR = \frac{12K}{C(C+1)} \sum_{j=1}^{C} \left( R_{\bullet j} - \frac{C+1}{2} \right)^2$$

$$= \left[ \frac{12}{KC(C+1)} \sum_{j=1}^{C} R_j^2 \right] - 3K(C+1).$$

The null hypothesis must be rejected in favor of the alternative for large values of the test statistic, that is at the significance level $\alpha$ when $FR_{obs} \geq FR_\alpha$, where $FR_\alpha$ is such that the probability of rejection when $H_0$ is true does not exceed $\alpha$. Alternatively the decision rule can be based on the probability that the test statistic assumes values greater than or equal to the observed one, that is the $p$-value, by comparing it with $\alpha$ as usual and rejecting $H_0$ when it is less than $\alpha$. Under the null hypothesis the test statistic asymptotically follows a chi-square distribution with $C - 1$ degrees of freedom. In the case of ties the mid-rank rule and a suitable correction to the test statistic must be applied (Hollander and Wolfe, 1999).

For example, in an acoustic experiment on listening conditions in classrooms, a possible goal could be to test whether the speech perception of the pupils can be affected by the type of noise at the significance level $\alpha = 0.01$. The response variable of the problem may be represented by the speech intelligibility (SI), that is the percentage of words correctly understood by the listeners in a suitable acoustic experiment (Prodi et al., 2010). The experiment was performed on 7 pupils. A female teacher's voice pronounced a sentence containing the target word with background noises proposed in turn together with the speech: 'activity in class', 'tapping' and 'external traffic' noises. The experimental data are shown in Table 3.5.

Table 3.5   Speech intelligibility in an acoustic experiment in classrooms of a primary school with different types of background noises.

| Pupil | Background noise | | |
|---|---|---|---|
| | Activity | Tapping | Traffic |
| 1 | 51.02 | 73.90 | 59.33 |
| 2 | 73.43 | 74.47 | 69.06 |
| 3 | 79.45 | 86.28 | 73.38 |
| 4 | 71.75 | 86.38 | 72.14 |
| 5 | 58.78 | 74.13 | 63.17 |
| 6 | 60.13 | 79.52 | 69.11 |
| 7 | 63.77 | 75.57 | 67.72 |

Here each block is represented by a single student ($K = 7$) and we are in the presence of $C = 3$ treatments (noises). To perform the described procedure for testing a significant effect of the type of noise on SI the following $R$ code may be applied:

```
> SI=matrix(c(51.02, 73.43, 79.45, 71.75, 58.78, 60.13, 63.77,
+ 73.90, 74.47, 86.28, 86.38, 74.13, 79.52, 75.57,
+ 59.33, 69.06, 73.38, 72.14, 63.17, 69.11, 67.72),
+ nrow = 7, byrow = FALSE, dimnames = list(1 : 7,
+ c("Activity", "Tapping", "Traffic")))
> friedman.test(SI)
```

The matrix of data with rows corresponding to blocks and columns corresponding to treatments, must be defined. Then the `friedman.test` function, which requires just the matrix in input, performs the rank based procedure. The following output is returned:

```
#######################################################################
# Friedman rank sum test
# data: SI
# Friedman chi-squared = 11.1429, df = 2, p-value = 0.003805
#######################################################################
```

The observed value of the test statistic is equal to 11.143. According to the tabled distribution reported in Hollander and Wolfe (1999), the upper-tail probability $Pr\{FR \geq 11.143\}$ with 7 blocks and 3 treatments, is 0.001. Hence the $p$-value 0.001 is less than $\alpha = 0.01$ and the null hypothesis must be rejected in favor of the hypothesis of non-null effect of the type of noise on the SI. Among the outputs of the program, the degrees of freedom and the $p$-value according to the asymptotic chi-square null distribution are also returned by the `friedman.test` function. In this case the asymptotic $p$-value 0.004 leads to the rejection of the null hypothesis too.

### 3.3.2 Permutation test for related samples

Another nonparametric solution for unreplicated block designs is the so-called permutation test for related samples. The typical problem for related samples concerns the case of repeated measures, when $n = K$ units are observed $C$ times, and time takes the role of symbolic treatment. This layout, when units play the role of blocks and the interest is in the treatment effects, may correspond to an unreplicated two-way ANOVA, where for each combination of block level and treatment level, just one replication is considered. No interaction is assumed between units and occasions of measurement (Pesarin and Salmaso, 2010).

Let us assume that the underlying unspecified distributions are non-degenerate and also that:

1. the $K$ profiles are independent;

2. the within block responses are homoscedastic and exchangeable with respect to treatments in the null hypothesis (note that this interpretation is appropriate

for most cases of repeated observations when there are no interaction effects between units and time);

3. the underlying model for fixed effects is additive, that is $Z_{kj} = \theta + \beta_k + \mu_j + \epsilon_{kj}$, with $k = 1, \ldots, K$ and $j = 1, \ldots, C$, where $\theta$ is a population constant (e.g., unknown mean or median), $\beta_k$ represents the $k$th block effect, $\mu_j$ is the unknown effect of the $j$th treatment and $\epsilon_{kj}$ are exchangeable random errors with zero mean or median with common distribution and standard deviation $\sigma_k$, which can vary with respect to blocks.

Unlike the Friedman test, the permutation solution relaxes the continuity assumption for the response variable and it can be applied in the presence of discrete variables. Furthermore, with the permutation test, also the block effect, ignored in the Friedman approach, can be tested (Pesarin, 2001).

To test the null hypothesis of equal treatment effects $H_0 : [\mu_1 = \cdots = \mu_C]$ against the alternative $H_1 : [\mu_1, \ldots, \mu_C$ not all equal] we can consider that in $H_0$ data within each block are exchangeable with respect to treatments, and that they are independent with respect to units. A suitable test statistic may reasonably have the same form as in the parametric solution

$$T_{RS}^* = \frac{\sum_{j=1}^{C} \left( \overline{X}_{\cdot j}^* - \overline{X}_{\cdot\cdot} \right)^2}{\sum_{k=1}^{K} \sum_{j=1}^{C} \left( X_{kj}^* - \overline{X}_{k\cdot} - \overline{X}_{\cdot j}^* + \overline{X}_{\cdot\cdot} \right)^2}$$

where $\overline{X}_{\cdot j}^* = \sum_k X_{kj}^* / K$, $j = 1, \ldots, C$, $\overline{X}_{k\cdot} = \sum_j X_{kj}^* / C$, $k = 1, \ldots, K$ and $\overline{X}_{\cdot\cdot} = \sum_{k,j} X_{kj}^* / (KC)$ are column, row and global means with respect to the $K \times C$ data matrix after a given permutation. Since the permutations are performed within blocks (rows) and with respect to the treatment levels (columns), the block means and the global mean are permutationally invariant, that is $\overline{X}_{k\cdot}^* = \overline{X}_{k\cdot}$ and $\overline{X}_{\cdot\cdot}^* = \overline{X}_{\cdot\cdot}$. The null hypothesis must be rejected for large values of $T_{RS}$, hence the $p$-value is the probability that $T_{RS}^*$ takes values greater than or equal to the observed value of $T_{RS}$.

Let us consider the same example discussed in Section 3.3.1 and solved with the Friedman test, the acoustic experiment on listening conditions in classrooms to test whether the SI, that is the percentage of words correctly understood by pupils, can be affected by the type of noise at the significance level $\alpha = 0.01$. $C = 3$ types of noise were taken into account in the experiment ('activity in class', 'tapping', and 'external traffic') and $K = 7$ pupils were involved in the tests (Table 3.5). The $R$ code to perform the permutation test requires the files two.way.rs.r and the usual t2p.r:

```
> source("twoWayRs.r")
> source("t2p.r")
> SI=matrix(c(51.02, 73.43, 79.45, 71.75, 58.78, 60.13, 63.77,
+ 73.90, 74.47, 86.28, 86.38, 74.13, 79.52, 75.57,
+ 59.33, 69.06, 73.38, 72.14, 63.17, 69.11, 67.72),
```

```
+ nrow = 7, byrow = FALSE, dimnames = list(1 : 7,
+ c("Activity", "Tapping", "Traffic")))
> two.way.rs(SI,B=10,000)
###########################################################################
# $p.value
# 0.000999001
###########################################################################
```

The function two.way.rs, included in the script twoWayRs.r and that requires the script t2p.r for the computation of the $p$-value, allows the described procedure to be performed and requires the $K \times C$ data matrix as input. The final $p$-value of the permutation test is 0.001, thus the null hypothesis must be rejected in favor of the hypothesis that the type of noise affects the SI.

### 3.3.3     The Page test for ordered alternatives

The nonparametric Page test is useful in the two-way ANOVA in a randomized complete block design when the alternative to the null hypothesis of no differences in the treatment effects is a specific ordering (increasing or decreasing) of the treatment effects (Page, 1963). Formally the hypotheses of the Page test are $H_0 : [\mu_1 = \cdots = \mu_C]$ against $H_1 : [\mu_1 \leq \mu_2 \leq \cdots \leq \mu_C$ with at least one strict inequality]. Hence the Page test is the rank-based methodological solution, corresponding to the Friedman test, for ordered alternatives.

The main assumptions for the application of the test are the following: data are realizations of random variables with CDF $F_{kj}$ ($k = 1, \ldots, K; j = 1, \ldots, C$), which do not differ in variability but possibly only in their locations and these differences in central tendencies, with respect to block or treatment effects, can be represented by an additive model, where no interactions between block and treatment factors are present. The first step of the procedure consists of labeling the treatments according to the expected order in the alternative hypothesis. For the computation of the observed value of the test statistic, the within block ranks considered in the Friedman test should be obtained. Hence let $X_{kj}$ be the observation related to the $k$th block and $j$th treatment. The within block rank is the rank of $X_{kj}$ in the set of observations of the $k$th block $\{X_{k1}, \ldots, X_{kC}\}$, that is $R_{kj} = \mathbb{R}_k(X_{kj})$. Let $R_j$ denote the sum of ranks for the $j$th treatment, that is $R_j = \sum_k R_{kj}$. The Page test statistic is given by

$$PG = \sum_{j=1}^{C} jR_j = R_1 + 2R_2 + \cdots + CR_C.$$

The null hypothesis must be rejected for large values of $PG$, that is when $PG_{obs} \geq PG_\alpha$, where $PG_\alpha$ is chosen to make the probability of wrongly rejecting $H_0$ in favor of $H_1$ equal to $\alpha$. The critical values, obtained by numerical approximation, are tabulated and available for different values of $C$ and $K$ (see Appendix C). According to the large sample approximation, the null distribution of the standardized Page test statistic

Table 3.6    Speech intelligibility in an acoustic experiment in classrooms of a primary school with different speech transmission quality levels.

| Pupil | Speech transmission quality | | | |
|---|---|---|---|---|
| | Bad | Poor | Fair | Good |
| 1 | 58.20 | 75.75 | 81.35 | 93.30 |
| 2 | 75.35 | 72.30 | 78.80 | 89.60 |
| 3 | 67.65 | 78.10 | 80.40 | 91.25 |
| 4 | 62.30 | 68.15 | 76.45 | 86.10 |
| 5 | 56.80 | 65.25 | 75.20 | 85.10 |
| 6 | 69.35 | 85.50 | 82.35 | 95.20 |

is normal (Randles and Wolfe, 1979). The mid-rank rule must be used in the case of ties.

Consider again the example of the acoustic experiment described in Section 3.3.2, where the response variable of the problem is the SI (Prodi *et al.*, 2010). The speech transmission index (STI) is an indicator of the speech transmission quality in classrooms that measures some physical characteristics of the classroom and expresses the ability of the classroom to carry across the characteristics of speech signals. It takes values from 0 to 1. To test whether the SI is an increasing function of the speech transmission quality (STQ), a new ordered categorical variable (*STQ*) may be defined according to the following rule (Bonnini *et al.*, 2014):

- if $STI \leq 0.3$ then $STQ = bad$;
- if $0.3 < STI \leq 0.4$ then $STQ = poor$;
- if $0.4 < STI \leq 0.6$ then $STQ = fair$;
- if $STI > 0.6$ then $STQ = good$.

The aim here is to test whether higher levels of STQ (treatment factor) correspond to greater values of SI. Data related to a sample of 6 students involved in this experiment are reported in Table 3.6

For the application of the Page test, an *R* function included in the crank library may be used (*2.15.3* or later *R* versions are needed). The syntax to create the matrix of data and perform the analysis, and obtain the final output are the following:

```
> library(crank)
> SI=matrix(c(0.58,0.75,0.67,0.62,0.56,0.69,
+ 0.75,0.72,0.78,0.68,0.65,0.85,
+ 0.81,0.78,0.80,0.76,0.75,0.82,
+ 0.93,0.89,0.91,0.86,0.85,0.95),
+ nrow=6,byrow=FALSE,dimnames=list(1:6,
+ c("Bad","Poor","Fair","Good")))
> R=matrix(1,nrow(SI),ncol(SI))
```

```
> for (k in 1:nrow(SI))
+ R[k,]=rank(SI[k,],ties.method="average")
> page.trend.test(R)
####################################################################
# Page test for ordered alternatives
# L = 178  p(table)  <=.001
####################################################################
```

After the definition of the matrix of data, the rank transformation for each block (pupil) is needed. The ranks are stored in the matrix $R$ (previously initialized) through the `for` cycle. The function `page.trend.test` performs the Page test and requires as input the $R$ matrix of ranks. The outputs consist of the observed value of the test statistic (L) and the $p$-value. In this case the observed value of the statistic is 178 (very large) and the corresponding $p$-value is less than 0.001, hence at the significance level $\alpha = 0.01$ we reject the null hypothesis in favor of the alternative of increasing ordering of SI with respect to STQ.

### 3.3.4   Permutation analysis of variance in the presence of two factors

In a $K \times J$ replicated factorial design, with fixed effects, balanced samples and homoscedasticity, two factors with $K$ and $J$ levels, respectively, are taken into account and, for each combination $(k, j)$ of levels, $n$ replications are considered. Let $X_{kji}$ denote the $i$th observation related to combination $(k, j)$, that is the value observed on the $i$th unit ($i$th replication) when factor 1 is at level $k$ and factor 2 is at level $j$ ($k = 1, \ldots, K; j = 1, \ldots, J$). Data are assumed to be realizations of random variables $Z_{kji}$ which behave according to the linear model:

$$Z_{kji} = \theta + \beta_k + \gamma_j + (\beta\gamma)_{kj} + \epsilon_{kji},$$

where $\beta_k$ and $\gamma_j$ are the main effects of factor 1 and factor 2, respectively, $(\beta\gamma)_{kj}$ are the interaction effects and $\epsilon_{kji}$ are exchangeable errors, with zero mean and unknown continuous distribution $F$ (Pesarin, 2001). Let us also assume the usual side-conditions:

$$\sum_{k=1}^{K} \beta_k = 0, \sum_{j=1}^{J} \gamma_j = 0, \sum_{k=1}^{K} (\beta\gamma)_{kj} = 0 \,\forall j, \sum_{j=1}^{J} (\beta\gamma)_{kj} = 0 \,\forall k.$$

Usually three possible testing problems are considered, separately from each other:

- $H_{0\beta} : [\beta_k = 0$ for every $k]$, against $H_{1\beta} : [\beta_k \neq 0$ for some $k]$ (significance of main effect of factor 1);

- $H_{0\gamma} : [\gamma_j = 0$ for every $j]$, against $H_{1\gamma} : [\gamma_j \neq 0$ for some $j]$ (significance of main effect of factor 2);

- $H_{0\beta\gamma} : [(\beta\gamma)_{kj} = 0$ for every $(k, j)]$ against $H_{1\beta\gamma} : [(\beta\gamma)_{kj} \neq 0$ for some $(k, j)]$ (significance of interaction effect).

Under the null hypothesis of one of the three problems, exchangeability holds only between specific data blocks (defined according to the $J \times K$ combinations of levels) and so only synchronized permutations are allowed (Pesarin and Salmaso, 2010). For example if we are interested in comparing two treatments of factor 1, corresponding to level $k_1$ and $k_2$ of the factor, with $1 \leq k_1 < k_2 \leq K$ we can permute data between blocks with the same level of factor 2, one with factor 1 at level $k_1$ and another with factor 1 at level $k_2$. In other words, by denoting with $(k, j)$ the block of data with factor 1 at level $k$ and factor 2 at level $j$, for every $j$ we can exchange data between $(k_1, j)$ and $(k_2, j)$ and the permutations between the $J$ couples of blocks must be synchronized. At any fixed level $j$ of factor 2, ${}^{\beta}T_{k_1 k_2 | j} = \sum_i X_{k_1 ji} - \sum_i X_{k_2 ji}$ is a suitable intermediate test statistic. Hence, considering that $K(K-1)/2$ pairwise comparisons are needed to test $H_{0\beta}$ against $H_{1\beta}$, a suitable test statistic for the permutation test on the main effect of factor 1 is:

$$T_{\beta}^* = \sum_{k_1 < k_2} \left[ \sum_j {}^{\beta}T_{k_1 k_2 | j}^* \right]^2 .$$

The hypothesis of null effect must be rejected for large values of the test statistic, that is when the observed value of the statistic is greater than or equal to the critical value (i.e., the $(1 - \alpha)$-quantile of the permutation distribution). Hence the $p$-value is $\Pr\{T_{\beta}^* \geq T_{\beta,obs}\}$ and, as usual, an equivalent decision rule consists in rejecting $H_0$ in favor of $H_1$ when the $p$-value $\leq \alpha$. A similar procedure can be applied to test for the main effect of factor 2 and for the interaction effect. The NPC methodology could be applied to obtain a global test by combining the partial tests based on pairwise comparisons between data blocks.

Let us suppose an interest in examining the effects of the factors 'oil brand' and 'oil change interval' on the oil impurities in a car engine. Let us consider the observed sample of $n = 12$ observations (replications) for each combination of brand (factor 1 with 2 levels: 'brand A' and 'brand B') and interval (factor 2 with 2 levels: '12 000 km' and '24 000 km'). The response is the so called 'purity'. Data are shown in Table 3.7.

The permutation test with constrained synchronized permutations for the two-way ANOVA design can be performed by the $R$ function CSP, that requires as input a vector of data y, representing the observed values of the response variable and a matrix of labels (categorical data) x, representing the factors level combinations. Each column of x corresponds to one factor. In the present example the first column of x denotes the levels of brand, whereas the second column reports the levels of distance. A further input of the function is the desired number of permutations. The CSP function provides as output the $p$-values related to the tests on the main effects and on the interaction effect. The $R$ code for the application of the procedure to the described problem is:

```
> source("CSP.r")
> data=read.csv("Motors.csv",header=TRUE,sep=";")
> data[,2]=ifelse(data[,2]=='12000',1,2)
```

Table 3.7   Purity of car engine, oil brand and traveled distance before the oil change in a two-way ANOVA experiment.

| Distance (km) | Brand A | Brand B | Distance (km) | Brand A | Brand B |
|---|---|---|---|---|---|
| 12 000 | 6.6 | 7.6 | 24 000 | 7 | 7.8 |
| 12 000 | 7.4 | 8.2 | 24 000 | 7.3 | 8.5 |
| 12 000 | 6.5 | 7 | 24 000 | 6 | 7.5 |
| 12 000 | 7.1 | 9.1 | 24 000 | 7.3 | 8.3 |
| 12 000 | 6.3 | 7.2 | 24 000 | 6.9 | 7.5 |
| 12 000 | 6.4 | 9.8 | 24 000 | 7.5 | 7.5 |
| 12 000 | 6.9 | 6.3 | 24 000 | 7 | 7 |
| 12 000 | 6.2 | 7.7 | 24 000 | 7.4 | 9.9 |
| 12 000 | 5.3 | 9 | 24 000 | 7.1 | 7.2 |
| 12 000 | 6.7 | 9.6 | 24 000 | 7 | 9.4 |
| 12 000 | 6.8 | 8.6 | 24 000 | 5.8 | 6 |
| 12 000 | 6.7 | 7.9 | 24 000 | 8 | 9.1 |

```
> data[,1]=ifelse(data[,1]=='A',1,2)
> x=cbind(data[,1],data[,2])
> y=data[,3]
> t=CSP(y,x,10000)
```

The source command allows the use of the script that includes the CSP function. The subsequent instructions are necessary for the recoding of the factor levels into integer numbers, the creation of y vector and x matrix and the performance of the test. To see the final output type:

```
> c(t$pa,t$pb,t$pab)
0.0002 0.6569 0.1546
```

Hence the results show that the only significant effect is the main effect of factor A, that is the main effect of brand ($p$-value= 0.000). The main effect of distance and the interaction effect between the two factors on the engine purity are not significant. Hence the evidence is that purity depends only on the oil brand.

## 3.4   Pairwise multiple comparisons

In the presence of $C$ samples drawn from $C$ populations we have discussed procedures to test the null hypothesis $H_0 : [\mu_1 = \cdots = \mu_C]$ against a variety of alternative hypotheses, where $C$ represents the number of treatments or levels of a given factor. In the case of rejection of $H_0$ the conclusions range from the general statement that there are some generic unspecified differences among the treatment effects (as in the Kruskal–Wallis test) to more specific and informative statements consisting of ordered or umbrella alternatives. However, none of these testing procedures investigate pair-specific comparisons between couples of samples. If we are also interested in comparing the effects of couples of treatments, we may use multiple comparison procedures. With multiple comparisons it is possible to compare three or more means, in pairs or combinations, of the same measurements (Westfall *et al.*, 1999). Do not confuse multiple comparison procedures with procedures related to multiple testing. Multiple testing procedures usually consider multiple measurements on the same statistical units giving rise to a multivariate testing problem.

A typical multiple comparison procedure, related to a one-way ANOVA design, consists of testing

$$H_0 : [\mu_1 = \cdots = \mu_C] \equiv \bigcap_{k<j} [\mu_k = \mu_j]$$

against

$$H_1 : [\mu_1, \ldots, \mu_C \text{ not all equal}] \equiv \bigcup_{k<j} [\mu_k \neq \mu_j],$$

where the null hypothesis states that each couple of treatments has equal effects and the alternative states that at least one couple of treatments has different effects. In the case of rejection of $H_0$, we might be interested in determining to which of the pairwise comparisons the significance of the test should be attributed.

## 3.4.1   Rank-based multiple comparisons for the Kruskal–Wallis test

For the Kruskall–Wallis test, a rank-based multiple comparison procedure to compare pairs of treatment effects has been proposed by Steel (1959, 1960, 1961), Dwas (1960), Fligner (1984, 1985) and Critchlow and Fligner (1991). For an in-depth discussion on this and other methods see Hollander and Wolfe (1999).

The $R$ package pgirmess includes the function kruskalmc that, when the Kruskal–Wallis test is significant, helps in determining which groups are different, performing pairwise comparisons with adjustment of the $p$-values for controlling the multiplicity of the test. Those pairs of samples for which the differences are higher than a critical value are considered statistically different at the given adjusted $\alpha$. As a matter of fact the significance levels of the partial tests related to the pairwise comparisons must be adjusted because if the probability of rejection of each partial test, when the partial null hypothesis is true, is equal to $\alpha$, the probability of rejection of the global null hypothesis $H_0$ is larger than $\alpha$. The higher the number of partial tests, that is the number of pairwise comparisons, the larger the global probability of type I error. Hence the control of the multiplicity of the test is needed and the type I error probability of each partial test must be consequently reduced (adjusted). Three types of multiple comparisons can be performed with this function: comparisons between treatments, 'one-tailed' and 'two-tailed' comparisons between treatments and control. The first factor level corresponds to the control.

For example let us apply the procedure to a pharmacological experiment on the effects of an antipyretic drug on the body temperature of guinea pigs. The data are stored in drugs.csv. In this one-way ANOVA problem four different types of drug ($T_1$, $T_2$, $T_3$ and $T_4$), with four different concentration levels of active ingredient, were administered to four distinct groups of animals (independent samples). The temperature change (in °C) after the administration of the drug was observed on each animal (see data in Table 3.2). The Kruskal–Wallis test returns a $p$-value equal to 0.002. The effectiveness of the drug is not the same for the four treatments because the null hypothesis of equal effects is rejected at the significance level $\alpha = 0.01$. According to the classical Bonferroni rule, the adjusted $p$-values could be equal to $\alpha/mc$, where $mc$ is the number of pairwise comparisons, 6 in this problem.

The $R$ command for the multiple comparisons procedure is kruskalmc(resp, categ, probs, cont), where resp and categ correspond to the response variable and the treatment factor, respectively, probs represents the adjusted $\alpha$ levels for the pairwise comparisons (default value 0.05) and cont takes value "one-tailed" or "two-tailed" according to the type of comparison with the control or NULL (default value) if pairwise comparisons between all the treatments, in the absence of a control group, are considered. Hence the $R$ commands for the performance of the multiple comparisons are:

```
> drugs=read.csv("drugs.csv",header=TRUE,sep=";")
> attach(drugs)
> temp=c(T1,T2,T3,T4)
> drug=factor(rep(1:4,each=15),labels=c("T1","T2","T3","T4"))
> kruskalmc(temp,drug,probs=0.01/6,cont=NULL)
```

The returned results are:

```
######################################################################
# Multiple comparison test after Kruskal--Wallis
# p.value: 0.001666667
# Comparisons
#          obs.dif   critical.dif difference
# T1-T2 26.766667     23.18175       TRUE
# T1-T3  0.200000     23.18175       FALSE
# T1-T4 32.633333     23.18175       TRUE
# T2-T3 26.566667     23.18175       TRUE
# T2-T4  5.866667     23.18175       FALSE
# T3-T4 32.433333     23.18175       TRUE
######################################################################
```

According to this output, the difference of the effects of $T_1$ and $T_3$ and the difference of the effects of $T_2$ and $T_4$ are not significant. All the other differences are significant.

## 3.4.2    Permutation tests for multiple comparisons

The NPC methodology allows multiple comparison procedures to be performed. The method consists of simultaneously testing the null hypotheses

$$H_{0(jk)} : \mu_j = \mu_k$$

against the alternative hypotheses

$$H_{1(jk)} : \mu_j \neq \mu_k$$

with $j, k = 1, \ldots, C, j < k$. The global problem of testing whether some of the effects are not equal can be solved performing $C \times (C - 1)/2$ permutation two-sample tests and then combining the significance level functions of the partial tests to obtain a univariate test statistic and compute the corresponding permutation $p$-value. A close testing procedure may be applied for adjusting the $p$-values of the partial tests and attributing the possible significance of the global test at some of the two-sample partial problems. The $R$ codes for this procedure are included in the files dataperm.r, perm_2samples.r, univpermtest.r, t2p.r, combine.r, pairwise_comparisons.r and FWEminP.r. The syntax is:

```
> source("dataperm.r")
> source("perm_2samples.r")
> source("t2p.r")
> source("combine.r")
> source("pairwise_comparisons.r")
> source("FWEminP.r")
> drugs=read.csv("drugs.csv",header=TRUE,sep=";")
> temp=c(drugs[,2],drugs[,3],drugs[,4],drugs[,5])
> drug=rep(1:4,each=15)
> dataset=cbind(drug,temp)
```

Hence data must be organized in stacked form, on a matrix with two columns, the first representing the treatment, the second representing the response variable, as follows:

```
      drug temp
[1,]   1    1.4
...   ...   ...
[16,]  2   -1.7
...   ...   ...
[31,]  3    0.2
...   ...   ...
[46,]  4   -3.9
...   ...   ...
[60,]  4   -1.7
```

The final command and the output are:

```
> perm.csample.pwc(data=dataset,B=10,000,fun="Fisher",alpha=0.01)
######################################################################
# $p.Glob
# 0.0000999
# $p.Partial
#      [,1] [,2]  [,3]   [,4]
# [1,] NA     0  0.879  0.000
# [2,] NA    NA  0.000  0.101
# [3,] NA    NA    NA   0.000
# [4,] NA    NA    NA     NA
######################################################################
```

where the function perm.csample.pwc requires as input: data (the dataset with the treatment in the first column and the response in the second), the number of permutations B, the combining function fun ('Fisher', 'Liptak' or 'Tippett') and the significance level $\alpha$.

For this example the global $p$-value is significant, at $\alpha = 0.01$ level, thus even the partial corrected $p$-values are computed. The significant differences of effects in the pairwise comparisons are the same as the rank-based multiple comparisons procedure for the Kruskall–Wallis test: $T_1 - T_2$, $T_1 - T_4$, $T_2 - T_3$ and $T_3 - T_4$.

## 3.5   Multivariate multisample tests

In this section the multivariate extension of the multisample problem is discussed. The problem representation is the trivial extension of the multivariate two-sample location problem. The dataset here consists of $n = n_1 + n_2 + \cdots + n_C$ observations from $C$ independent $q$-variate populations, where $n_j$ is the size of the $j$th sample. The $i$th multidimensional observation in the $j$th group is denoted by $X_{ji} = \{X_{ji1}, \ldots, X_{jiq}\}$. A rank based and a permutation solution are described.

### 3.5.1   A multivariate multisample rank-based test

Let $F_j(x)$ be the cumulative distribution function, belonging to the class of all continuous distribution functions, for the $j$th population with $j = 1, \ldots, C$. The hypotheses are

$$H_0 : F_1(x) = \cdots = F_C(x) = F(x) \; \forall x \in \mathcal{R}^q$$

against the alternative

$$H_1 : F_j(x) \neq F(x) \text{ for at least one } j.$$

For the translation-type alternatives, we let

$$F_j(x) = F(x + \delta_j)$$

for $j = 1, \ldots, C$, $x \in \mathcal{R}^q$ and $\delta_j \in \mathcal{R}^q$. Hence the null hypothesis can be written as

$$H_0 : \delta_1 = \cdots = \delta_C = 0$$

against the alternative

$$H_1 : \delta_j \neq 0 \text{ for some } j.$$

Let $R_{jih}$ be the rank of $X_{jih}$ in the set $\{X_{11h}, \ldots, X_{1n_1h}, \ldots, X_{C1h}, \ldots, X_{Cn_Ch}\}$ for $h = 1, \ldots, q$. Let $\mathbf{R}_h$ denote the $n$-dimensional vector of ranks related to the $h$th variable and consider the $n \times q$ matrix $\mathbf{R} = [\mathbf{R}_1, \ldots, \mathbf{R}_q]$. Under $H_0$ the distribution of the matrix of ranks conditional to the observed matrix $\mathbf{R}$ over the set $S(\mathbf{R}^*)$ of all the possible permutations of the rows $\mathbf{R}^*$ is uniform and the probability of observing one of the $n! = n \cdot (n-1) \cdot \ldots \cdot 1$, possible realizations is $1/n!$.

A general class of rank scores can be defined as follows:

$$E^{(h)}(R) = g_h\left(\frac{R}{n+1}\right),$$

with $h = 1, \ldots, q$ and $1 \leq R \leq n$. Hence the matrix $\mathbf{R}$ can be replaced by $E = [\mathbf{E}_1, \ldots, \mathbf{E}_q]$ where

$$\mathbf{E}_h = \left[E^{(h)}(R_{11h}), \ldots, E^{(h)}(R_{1n_1h}), \ldots, E^{(h)}(R_{C1h}), \ldots, E^{(h)}(R_{Cn_Ch})\right]'$$

with $h = 1, \ldots, q$. The average rank scores can be computed as:

$$\overline{T}_{j\bullet h} = \frac{\sum_{i=1}^{n_j} E^{(h)}(R_{jih})}{n_j}.$$

Under $H_0$ the average rank scores should be close to the total mean scores $\overline{T}_{\bullet\bullet h} = (n_1\overline{T}_{1\bullet h} + \cdots + n_C\overline{T}_{C\bullet h})/n$ and the contrasts $(\overline{T}_{j\bullet h} - \overline{T}_{\bullet\bullet h})$ should stochastically be close to zero for $j = 1, \ldots, C$ and $h = 1, \ldots, q$. In the presence of $C > 2$ groups a suitable test statistic for this problem might be

$$L = \sum_{j=1}^{C} n_j[\overline{T}_j - \overline{T}]'V^{-1}[\overline{T}_j - \overline{T}],$$

where $\overline{T}_j = [\overline{T}_{j\bullet 1}, \ldots, \overline{T}_{j\bullet q}]'$, $\overline{T} = [\overline{T}_{\bullet\bullet 1}, \ldots, \overline{T}_{\bullet\bullet q}]'$ and $V$ is the permutation covariance matrix of the contrasts $\overline{T}_j - \overline{T}$ under $H_0$. According to the type of scores, different tests can be performed. For the rank sum test the score is $E^{(h)}(R) = R/(n+1)$. Statistic $L$ is used in the parametric Hotelling's test which assumes normality. Hence it is a suitable test statistic under normality. When the multivariate distribution is not normal, this statistic may not be a valid choice.

Consider an example from agriculture concerning reforestation. Let us assume $C = 3$ samples of different types of tree, of sizes $n_1 = 53$, $n_2 = 51$ and $n_3 = 48$, respectively. For each tree, height and diameter were measured. We wish to test whether the three different species have the same diameter and height. To perform a multisample location test based on marginal ranks, we can use the function rank.ctest(formula) of the package ICSNP. In this, the input is the formula X~g where X is the numerical data matrix of the multivariate response and g a factor with $C$ levels. Since the dataset has $n = n_1 + n_2 + n_3 = 152$ rows, the table of the whole dataset (stored in the file trees.csv) is not shown. The $R$ code for the analysis is:

```
> library(ICSNP)
> data=read.csv("trees.csv",header=TRUE,sep=";")
> X=as.matrix(data[,2:3])
> g=rep(c(1,2,3),c(53,51,48))
> rank.ctest(X g,scores="rank")
```

The output of the analysis is:

```
############################################################################
# Marginal C sample Rank Sum Test
#
# data X by g
# T = 80.2306, df=4, p-value < 2.2e-16
############################################################################
```

The value of the test statistic, the degrees of freedom of the approximated chi-square distribution for the statistic and the $p$-value are shown. Note that the $p$-value is almost 0 and leads to the rejection of the null hypothesis of no difference between the 3 groups, in favor of the alternative that they differ in the location for at least one variable.

Consider now an example of customer satisfaction. Here $C = 4$ hotels were evaluated (each of them by a sample of $n = 10$ customers) with respect to $q = 3$ variables: cleanliness, courtesy and price. The data consist of rates from 0 (minimum satisfaction) to 100 (maximum satisfaction) and are shown in Table 3.8. We wish to test whether the customer satisfaction for the four hotels is equal or not. The $R$ commands for the analysis and the results are:

```
> library(ICSNP)
> data=read.csv("hotel.csv",header=TRUE,sep=";")
> X=as.matrix(data[,2:4])
> g=rep(c(1,2,3,4),each=10)
> rank.ctest(X g,scores="rank")
############################################################################
# Marginal C sample Rank Sum Test
#
# data X by g
# T = 27.0768, df=9, p-value = 0.001358
############################################################################
```

Table 3.8 Customer satisfaction rates (from 0 to 100) on cleanliness, courtesy and price for four different hotels.

| Hotel | Cleanliness | Courtesy | Price | Hotel | Cleanliness | Courtesy | Price |
|---|---|---|---|---|---|---|---|
| Bellevue | 49 | 39 | 61 | Star | 9 | 55 | 42 |
| Bellevue | 33 | 81 | 90 | Star | 22 | 26 | 35 |
| Bellevue | 55 | 69 | 47 | Star | 65 | 23 | 33 |
| Bellevue | 28 | 27 | 56 | Star | 40 | 62 | 16 |
| Bellevue | 60 | 8 | 49 | Star | 17 | 18 | 30 |
| Bellevue | 57 | 16 | 62 | Star | 1 | 49 | 26 |
| Bellevue | 25 | 45 | 31 | Star | 45 | 61 | 38 |
| Bellevue | 71 | 22 | 55 | Star | 5 | 31 | 52 |
| Bellevue | 43 | 53 | 34 | Star | 54 | 56 | 23 |
| Bellevue | 16 | 10 | 48 | Star | 44 | 36 | 8 |
| Italy | 48 | 74 | 18 | Dante | 46 | 47 | 66 |
| Italy | 89 | 16 | 80 | Dante | 24 | 52 | 10 |
| Italy | 68 | 48 | 81 | Dante | 13 | 37 | 37 |
| Italy | 41 | 70 | 95 | Dante | 30 | 51 | 21 |
| Italy | 72 | 67 | 36 | Dante | 63 | 21 | 46 |
| Italy | 58 | 73 | 60 | Dante | 38 | 50 | 50 |
| Italy | 75 | 60 | 92 | Dante | 42 | 29 | 51 |
| Italy | 53 | 66 | 73 | Dante | 76 | 32 | 44 |
| Italy | 79 | 80 | 27 | Dante | 19 | 15 | 45 |
| Italy | 64 | 34 | 100 | Dante | 20 | 59 | 32 |

Hence also in this case the $p$-value is very low and the null hypothesis of equal customer satisfaction for the four hotels must be rejected.

### 3.5.2 A multivariate multisample permutation test

An extension of the permutation one-way ANOVA procedure for the case of multivariate responses is now presented. The methodological solution belongs to the family of NPC procedures. Let us refer to a one-way MANOVA (i.e., multivariate analysis of variance) design. The $C$ samples are presumed to be related to $C$ levels of a treatment and data within each sample are supposed to be realizations of i.i.d. random variables following the distribution $F_j$, $j = 1, \ldots, C$ (in place of independence, exchangeability may generally suffice). As usual the null hypothesis refers to equality of the $C$ multivariate distributions $F_1, \ldots, F_C$. By assuming that the $n_j$ observations of the $j$th sample are realizations of the $q$-dimensional random variable $\mathbf{Z}_j$, the null hypothesis is

$$H_0 : \{F_1 = \cdots = F_C\} \equiv \left\{ \mathbf{Z}_1 \overset{d}{=} \cdots \overset{d}{=} \mathbf{Z}_C \right\}.$$

The null hypothesis can be broken down into a finite set of $q$ sub-hypotheses,

$$H_0 : \cap_{h=1}^{q} \left\{ Z_{1h} \overset{d}{=} \cdots \overset{d}{=} Z_{Ch} \right\},$$

each corresponding to a one-way univariate ANOVA, where $[Z_{j1}, \ldots, Z_{jq}]^| = \mathbf{Z}_j$, $j = 1, \ldots, C$. Therefore, $H_0$ is true if all the $q$ partial hypotheses are jointly true. In this sense, $H_0$ is also called the global or overall null hypothesis. $H_0$ implies that the $q$-dimensional data vectors in $\mathbf{X}$ are exchangeable with respect to the $C$ groups. The alternative hypothesis states that at least one of the $q$ null sub-hypotheses is not true. Hence, the alternative may be represented by the union of $q$ sub-alternatives,

$$H_1 : \cup_{h=1}^{q} \left\{ Z_{jh} \overset{d}{\neq} Z_{kh} \text{ for some } j, k = 1, \ldots, C \right\},$$

stating that $H_1$ is true when at least one sub-alternative is true. In this context, $H_1$ is called the global or overall alternative. Each partial null and alternative hypothesis can be further broken down into $C(C - 1)/2$ sub-hypotheses if you want to tackle the problem in terms of pairwise sample comparisons. Hence this problem can be addressed by performing dependent permutation tests and combining the results by means of the NPC methodology.

Consider again the bivariate example related to reforestation introduced in the previous subsection and let us apply the multivariate permutation procedure just described, using the pairwise comparisons method for the univariate ANOVA problem. In this case $2 \times 3 \times (3 - 1)/2 = 6$ partial tests are considered. The $R$ code is:

```
> source("dataperm.r")
> source("umultiaspect.r")
> source("t2p.r")
```

```
> source("combine.r")
> source("pairwise_comparisons.r")
> source("FWEminP.r")
> data=read.csv("trees.csv",header=TRUE,sep=";")
> data[,1]=rep(c(1,2,3),c(53,51,48))
> perm.csample.pwc(data,B=10,000,fun="Fisher",alpha=0.05,type.perm=
  "asynchro")
```

To perform the analysis the command pairwise.csample.pwc(data,B,fun, alpha,type.perm) may be used, where the inputs data, B, fun, alpha and type.perm correspond to the dataset, the number of permutations, the combining function, the significance level, and the type of permutations, respectively. Note that in this example we have unbalanced samples and we have to specify that the permutations must be asynchronised. $R$ returns the following output:

```
###############################################################
#
# $p.Glob
# 0.0000999001
# $p.Partial
#      [,1]     [,2] [,3]
# [1,] NA    0.1748    0
# [2,] NA       NA     0
# [3,] NA       NA    NA
#
###############################################################
```

The global $p$-value is less than $\alpha$ and the null hypothesis of equal height and diameter between the three species must be rejected. According to the partial adjusted $p$-values, the rejection of the null hypothesis should be attributed to the third species 'Douglasia', significantly different from the others.

Consider now the example related to customer satisfaction of $C = 4$ hotels evaluated with respect to 3 variables (cleanliness, courtesy and price) by $n = 10$ customers (Table 3.8). Similarly to the previous example the $R$ code and the final output are:

```
> source("dataperm.r")
> source("umultiaspect.r")
> source("t2p.r")
> source("combine.r")
> source("pairwise_comparisons.r")
> source("FWEminP.r")
> data=read.csv("hotel.csv",header=TRUE,sep=";")
> data[,1]=rep(c(1:4),each=10)
> perm.csample.pwc(data,B=10,000,fun="Fisher",alpha=0.05)
```

```
####################################################################
#
# $p.Glob
# 0.00009999
# $p.Partial
#      [,1]    [,2]    [,3]    [,4]
#[1,] NA   0.0139 0.0139 0.5232
#[2,] NA       NA 0.0000 0.0000
#[3,] NA       NA     NA 0.5232
#[4,] NA       NA     NA     NA
#
####################################################################
```

In this example the global $p$-value is less than 0.0001 and the null hypothesis of equal satisfaction for the four hotels is rejected in favor of the alternative. At the significance level $\alpha = 0.05$ the customer satisfaction of hotels 1 (*Bellevue*) and 4 (*Dante*) and that of hotel 3 (*Star*) and 4 (*Dante*) are not significantly different.

# References

Alhakim, A. and Hooper, W. (2008) A non-parametric test for several independent samples. Journal of Nonparametric Statistics, 20, 253–261.

Bagdonavicius, V., Kruopis, J. and Nikulin M.S. (2011) Non-parametric Tests for Complete Data. John Wiley & Sons, Ltd.

Basso, D. and Salmaso, L. (2011) A permutation test for umbrella alternatives. Statistics and Computing, 21, 45–54.

Bonnini, S., Prodi, N., Salmaso, L. and Visentin, C. (2014) Permutation approaches for stochastic ordering. Communication in Statistics: Theory and Methods, in press.

Critchlow, D.E. and Fligner, M.A. (1991) On distribution-free multiple comparisons in the one-way analysis of variance. Communication in Statistics: Theory and Methods, 20, 127–139.

Dwass, M. (1960) Some k-sample rank-order tests. In: Olkin, I., Ghurye, S.G., Hoeffding, H., Madow, W.G. and Mann, H.B. (eds) Contributions to Probability and Statistics. Stanford University Press, pp. 198–202.

Fligner, M.A. (1984) A note on two-sided distribution-free treatment versus control multiple comparisons. Journal of the American Statistical Association, 79, 208–211.

Fligner, M.A. (1985) Pairwise versus joint ranking: Another look at the Kruskal–Wallis statistic. Biometrika, 72, 705–709.

Friedman, M. (1937) The use of ranks to avoid the assumption of normality implicit in the analysis of variance. Journal of the American Statistical Association, 32, 675–701.

Hollander, M. and Wolfe, D.A. (1999) Nonparametric Statistical Methods, 2nd edn. John Wiley & Sons, Ltd.

Klotz, J. and Teng, J. (1977) One-way layout for counts and the exact enumeration of the Kruskal–Wallis H distribution with ties. Journal of the American Statistical Association, 72, 165–169.

Kruskal, W.H. and Wallis, W.A. (1952) Use of ranks in one-criterion variance analysis. Journal of the American Statistical Association, 47, 583–621.

Kvam, P.H. and Vidakovic, B. (2007) Nonparametric Statistics with Applications to Science and Engineering. John Wiley & Sons, Ltd.

Mack, G.A. and Wolfe, D.A. (1981) K-sample rank-tests for umbrella alternatives. Journal of the American Statistical Association, 76, 175–181.

Meyer, J.P. and Seaman, M.A. (2013) A comparison of the exact kruskal–wallis distribution to asymptotic approximations for all sample sizes up to 105. The Journal of Experimental Education, 81, 139–156.

Page, E.B. (1963) Ordered hypotheses for multiple treatments: a significance test for linear ranks. Journal of the American Statistical Association, 58, 216–230.

Pesarin, F. and Salmaso, L. (2010) Permutation Tests for Complex Data: Theory, Applications and Software. John Wiley & Sons, Ltd.

Prodi, N., Visentin, C. and Farnetani, A. (2010) Intelligibility, listening difficulty and listening efficiency in auralized classrooms. Journal of Acoustical Society of America, 128, 172–181.

Randles, R.H. and Wolfe, D.A. (1979) Introduction to the Theory of Nonparametric Statistics. John Wiley & Sons, Ltd.

Rust, S.W. and Fligner, M.A. (1984) A modification to the Kruskal–Wallis statistic for the generalized Behrns–Fisher problem. Communication in Statistics: Theory and Methods, 13, 2013–2028.

Steel, R.G.D. (1959) A multiple comparison rank sum test. Treatment versus control. Biometrics, 15, 560–572.

Steel, R.G.D. (1960) A rank sum test for comparing all pairs of treatments. Technometrics, 2, 197–207.

Steel, R.G.D. (1961) Some rank sum multiple comparisons tests. Biometrics, 17, 539–552.

Westfall, P. H., Tobias, R. D., Rom, D., Wolfinger, R. D. and Hochberg, Y. (1999) Multiple Comparisons and Multiple Tests using SAS. SAS Institute Inc.

# 4

# Paired samples and repeated measures

## 4.1 Introduction

In this chapter, location problems in the presence of two or more related samples are presented. In related sample designs the compared samples are not independent in the sense that the $i$th statistical unit (observation) of one sample is associated with the $i$th statistical unit (observation) of each of the other samples. For example, in the two-sample test, a typical problem with related samples is the so called case-control clinical study. In this type of study, a group of cases (patients with a specific disease or risk factor) are compared with a group of controls (people who do not have the disease or the risk factor). To reduce the effect on the results of possible confounding factors, the controls are selected with the matching technique: that is for each case there is a control who is similar according to one or more variables (age, gender, ethnic group, etc.). The typical multisample problem with related samples concerns the case of repeated measure designs. In these problems data are collected on the same $n$ statistical units in $C > 2$ different occasions, such that each time $j$ corresponds to $n$ sample data observed on the same $n$ units. Certainly in problems with related samples only the balanced case is possible.

By assuming that $n$ is the sample size and $C$ the number of dependent samples, the dataset is represented by $\{X_{ji}; i = 1, \ldots, n; j = 1, \ldots, C\}$, where $X_{1i}, X_{2i}, \ldots, X_{Ci}$ represent the $C$ data observed on the $i$th unit. Section 4.2 is dedicated to two-sample problems with paired data. In Section 4.3 tests for repeated measures are discussed.

*Nonparametric Hypothesis Testing: Rank and Permutation Methods with Applications in R,*
First Edition. Stefano Bonnini, Livio Corain, Marco Marozzi and Luigi Salmaso.
© 2014 John Wiley & Sons, Ltd. Published 2014 by John Wiley & Sons, Ltd.
Companion website: http://www.wiley.com/go/hypothesis_testing

## 4.2    Two-sample problems with paired data

The representation of the two-sample location problem for paired data (or paired samples) is similar to that for independent samples. The difference lies in the assumption that the observations of the two samples are realizations of two non-independent variables. In other words, the data are pairs of replications and each pair can be considered 'pre-treatment' and 'post-treatment' observations with reference to the symbolic (or real) treatment that defines the two samples. The goal of the study consists of testing the significance of a shift in location due to the application of the 'treatment'. The first assumption is that the subjects/objects are 'homogeneous' with respect to the most important experimental conditions, the so called covariates. The set of data pairs $(X_{1i}, X_{2i})$, $i = 1, \ldots, n$ may be viewed as a random sample of $n$ pairs from the bivariate random variable $\mathbf{Z} = (Z_1, Z_2)$ with $Z_1$ and $Z_2$ non-independent random variables with cumulative distribution function (CDF) $F_1$ and $F_2$, respectively. The hypotheses of the two-sided problem are

$$H_0 : \left\{ Z_1 \overset{d}{=} Z_2 \right\} \equiv \{ F_1(x) = F_2(x) \ \forall x \in \mathcal{R} \}$$

against the general two-sided alternative

$$H_1 : \left\{ Z_1 \overset{d}{\neq} Z_2 \right\} \equiv \{ F_1(x) \neq F_2(x) \text{ for some } x \in \mathcal{R} \}.$$

Of course, similarly to the location problem for independent samples, one-sided problems can also be defined and solved.

### 4.2.1    The Wilcoxon signed rank test

Consider the independent individual differences $D_i = X_{2i} - X_{1i}$, $i = 1, \ldots, n$, let us assume that $F_1$ and $F_2$ are completely unknown, symmetric about the median and continuous, so that ties in the observations are assumed to occur with probability zero. A suitable solution based on ranks evaluated on absolute values of differences is provided by the Wilcoxon signed rank test (Wilcoxon, 1945). In this framework we do not need to assume $\mathbb{E}(X)$ to be finite (Pesarin, 2001). To compute the Wilcoxon signed rank statistic $T^+$, sort the absolute values of the differences $|D_1|, \ldots, |D_n|$ from the smallest to the largest. Let $R_i$ denote the rank of $|D_i|$, $i = 1, \ldots, n$, that is $R_i = \mathbb{R}(|D_i|)$. Define the indicator variables $W_i$, $i = 1, \ldots, n$ where $W_i = 1$ if $D_i > 0$ and 0 otherwise, and obtain the $n$ products $R_1 W_1, \ldots, R_n W_n$. The product is known as the positive signed rank of $D_i$. It takes value zero if $D_i$ is negative and the rank of $|D_i|$ when $D_i$ is positive. The Wilcoxon signed rank statistic $T^+$ is the sum of the positive signed ranks,

$$T^+ = \sum_{i=1}^{n} R_i W_i$$

(Hollander and Wolfe, 1999). The $p$-value of the test can be computed according to the distribution of $T^+$ under $H_0$, that is considering all the possible values of $T^+$ which depend on all the $2^n$ possible sequences of 0 and 1 for $(W_1, \ldots, W_n)$. Alternatively we can follow the method of the rejection region, hence, at the significance level $\alpha$:

- For the two-sided test, we reject $H_0$ if $T^+ \geq t_{\alpha/2}$ or $T^+ \leq \frac{n(n+1)}{2} - t_{\alpha/2}$, with $t_{\alpha/2}$ such that the type I error probability is equal to $\alpha$.

- When the alternative is upper tailed one-sided $(H_1 : Z_2 \overset{d}{>} Z_1)$ we reject the null hypothesis if $T^+ \geq t_\alpha$;

- When the alternative is lower-tailed one-sided $(H_1 : Z_2 \overset{d}{<} Z_1)$ we reject the null hypothesis if $T^+ \leq \frac{n(n+1)}{2} - t_\alpha$.

Values of $t_\alpha$ are tabulated. The large-sample approximation is based on the asymptotic normality of the standardized version of $T^+$. When $H_0$ is true, the expected value and variance of $T^+$ are

$$\mathbb{E}_0(T^+) = \frac{n(n+1)}{4}$$

and

$$\mathbb{V}_0(T^+) = \frac{n(n+1)(2n+1)}{24},$$

respectively. The standardized version of $T^+$ is then

$$\widetilde{T}^+ = \frac{T^+ - \mathbb{E}_0(T^+)}{\sqrt{\mathbb{V}_0(T^+)}}.$$

When $H_0$ is true, $\widetilde{T}^+$ is asymptotically $\mathcal{N}(0,1)$ distributed.

Consider for example a treatment for body weight loss. The weights of $n = 12$ subjects before and after the treatment were considered to test whether this treatment is effective at the significance level $\alpha = 0.05$. Thus, the observations refer to the same subjects in two different moments. The expected effect is that the treatment leads to a reduction of the weight. By denoting with $Weight_{pre}$ and $Weight_{post}$ the random variables representing the weight of a patient before and after the treatment respectively, the hypotheses are

$$H_0 : Weight_{pre} \overset{d}{=} Weight_{post}$$

against

$$H_1 : Weight_{post} \overset{d}{<} Weight_{pre}.$$

Hence small values of the test statistic lead to the rejection of $H_0$ in favor of $H_1$.

The observed data are shown in Table 4.1.

The basic package of $R$ provides the function `wilcox.test(x,y,alternative, mu,paired,exact)` that performs one-sample or two-sample Wilcoxon tests on vectors of data. Among the required inputs, the vectors of the observations (x and y) are needed. If only x is given and y=NULL, the one-sample test is performed; if both x

Table 4.1   Body weights (kg) of patients before and after a treatment for weight loss.

| Patient | Pre-treatment | Post-treatment |
|---------|---------------|----------------|
| 1  | 117.48 | 90.20  |
| 2  | 83.90  | 87.46  |
| 3  | 92.29  | 87.15  |
| 4  | 103.89 | 76.16  |
| 5  | 94.94  | 89.19  |
| 6  | 121.87 | 93.94  |
| 7  | 105.71 | 94.69  |
| 8  | 87.07  | 81.57  |
| 9  | 100.00 | 98.39  |
| 10 | 91.30  | 99.07  |
| 11 | 98.99  | 103.49 |
| 12 | 96.78  | 100.98 |

and y are given and `paired=TRUE`, the Wilcoxon signed rank test is performed. The argument `alternative` must be equal to one of `"two-sided"` (default), `"greater"` or `"less"` according to the alternative hypothesis. The value of `mu` denotes the hypothesized difference x-y under $H_0$. Finally the logical parameter `exact` must take value TRUE if the exact $p$-value of the test is required instead of the asymptotic one. By default, an exact $p$-value is computed when less than 50 sample values are considered without ties. The $R$ code for the analysis and the results are:

```
> w_pre=c(117.48,83.90,92.29,103.89,94.94,121.87,
+ 105.71,87.07,100.00,91.30,98.99,96.78)
> w_post=c(90.20,87.46,87.15,76.16,89.19,93.94,
+ 94.69,81.57,98.39,99.07,103.49,100.98)
> wilcox.test(x=w_post,y=w_pre,alternative="less",mu=0,paired=TRUE,
   exact=TRUE)
####################################################################
# Wilcoxon signed rank test
#
# data: w_post and w_pre
# V=17 , p-value =0.04614
#alternative hypothesis: true location shift is less than 0
####################################################################
```

The function `wilcox.test` returns the value of the test statistic (V) and the $p$-value for the test (p-value). Note that the related $p$-value here is equal to 0.046 < 0.05 and leads to the rejection of the null hypothesis of no effect of the treatment in favor of the alternative hypothesis of reduction of body weight at the significance level $\alpha = 0.05$.

Table 4.2   Heart rate recovery of runners who do or
do not monitor their heartbeat (matched samples).

| Runners | No monitoring | Monitoring |
|---------|---------------|------------|
| 1 | 33 | 36 |
| 2 | 43 | 41 |
| 3 | 50 | 45 |
| 4 | 48 | 47 |
| 5 | 36 | 35 |
| 6 | 31 | 37 |
| 7 | 41 | 32 |
| 8 | 45 | 49 |

Table 4.2 shows data related to another application involving an experiment to evaluate whether the monitoring of the heartbeat adversely affects the recovery of a runner, that is it induces a lower reduction in heart rate after 1 min from the end of a race. Heart rate is computed as heartbeats per minute. Heart rate recovery (HRR) is equal to the difference between the heart rate immediately after exercise and the heart rate after 1 min. The larger the difference, the greater the health and fitness of the athlete. The samples are two groups of long distance runners matched in terms of age and personal best performance in a marathon: each athlete with a specific age and a best performance in the group of runners who do not monitor their heartbeat is matched with a similar athlete of the same age and with the same best performance in the group of runners who do monitor their heartbeat. The goal of the analysis is to test whether HRR is greater for athletes who do not monitor their heartbeat at the significance level $\alpha = 0.10$. Formally

$$H_0 : HRR_{no} \overset{d}{=} HRR_{yes}$$

against

$$H_1 : HRR_{no} \overset{d}{>} HRR_{yes},$$

where $HRR_{no}$ and $HRR_{yes}$ denote the HRR of runners who do not monitor their heartbeat and those that do respectively.

The R commands for the applications of the Wilcoxon signed rank test and the results are:

```
> HRR_no=c(33,43,50,48,36,31,41,45)
> HRR_yes=c(36,41,45,47,35,37,32,49)
> wilcox.test(x=HRR_no,y=HRR_yes,paired=TRUE,alternative="greater")
######################################################################
# Wilcoxon signed rank test
#
# data: HRR_no and HRR_yes
```

```
# V=20 , p-value =0.4167
#alternative hypothesis: true location shift is greater than 0
###################################################################
```

The $p$-value is equal to 0.417, hence the null hypothesis of no difference between the two groups cannot be rejected. In this case a warning message occurs to inform the user that the solution is not exact because the approximation is needed for the presence of ties.

Another interesting example of a paired data problem concerns a study of industrial policy where the policy efforts in several prefectures in southern China, specifically in the region of Guangdong, in 2001 and in 2008 are considered. The goal of the statistical analysis is to test whether the industrial policy efforts during the period 2001–2008 tend to increase, that is whether the efforts in 2008 were greater than in 2001 (Barbieri $et\ al.$, 2012). The response variable is a composite index that measures the level of industrial policy efforts performed in that period. The index sum ups information about the following variables:

1. *Specialized towns*: the variable indicates, for each prefecture and for each year, the number of towns officially recognized as specialized by the Department of Science and Technology of Guangdong Province, weighted for the number of years passed since each town obtained the official recognition.

2. *Development zones*: the variable summarizes for each prefecture the number of development zones weighted for the number of years since each of them was established by the national, provincial or local government.

3. *Innovation centers*: it identifies the number of innovation platforms, productivity promotion centers, etc., built by the provincial or the local government weighted by the number of years since the establishment of each center.

4. *Expenditure/revenues ratio*: it is the ratio between the prefecture's government expenditures and revenues.

5. *R&D expenditures*: it refers to the expenditures for research and development of the public research institutions. The index has been computed for each of the 21 prefectures applying the nonparametric ranking combination (Lago and Pesarin, 2000; Arboretti $et\ al.$, 2007) and it has been normalized such that values close to 0 indicate a low industrial effort, while values close to 1 indicate a high industrial effort with respect to the other prefectures. Since the goal of the study is to test whether the policy effort in 2008 is significantly greater than the policy effort in 2001 at the significance level $\alpha = 0.05$, formally the hypotheses are

$$H_0 : \textit{Effort}_{2008} \overset{d}{=} \textit{Effort}_{2001}$$

against

$$H_1 : \textit{Effort}_{2008} \overset{d}{>} \textit{Effort}_{2001}.$$

Data are stored in the file "china.csv" and shown in Table 4.3.

Table 4.3   Index of industrial policy effort for prefectures
of Guang dong (China) in the period 2001–2008.

| Prefecture | 2001 | 2008 |
|---|---|---|
| Chaozhou | 0.0355 | 0.0321 |
| Dongguan | 0.0214 | 0.0138 |
| Foshan | 0.3453 | 0.4881 |
| Guangzhou | 0.9925 | 0.9927 |
| Heyuan | 0.1406 | 0.1173 |
| Huizhou | 0.0166 | 0.0131 |
| Jiangmen | 0.0136 | 0.0389 |
| Jieyang | 0.0075 | 0.0349 |
| Maoming | 0.0090 | 0.0230 |
| Meizhou | 0.0206 | 0.0503 |
| Qingyuan | 0.0190 | 0.0132 |
| Shantou | 0.0168 | 0.0394 |
| Shanwei | 0.0138 | 0.0208 |
| Shaoguan | 0.0085 | 0.0084 |
| Shenzhen | 0.0462 | 0.0527 |
| Yangjiang | 0.0113 | 0.0291 |
| Yunfu | 0.0151 | 0.0358 |
| Zhanjiang | 0.0232 | 0.0348 |
| Zhaoqing | 0.0110 | 0.0183 |
| Zhongshan | 0.0420 | 0.0232 |
| Zhuhai | 0.0123 | 0.0073 |

For this testing problem the $R$ code and the results are:

```
> china=read.csv("china.csv",header=TRUE,sep=";")
> Effort01=china[,2]
> Effort08=china[,3]
> wilcox.test(x=Effort08,y=Effort01,paired=TRUE,alternative="greater")
######################################################################
# Wilcoxon signed rank test
#
# data:  Effort08 and Effort01
# V = 60, p-value = 0.02735
# alternative hypothesis: true location shift is greater than 0
######################################################################
```

Note that, now we have specified `alternative="greater"` because we want to test
whether the response takes greater values in 2008. The $p$-value is equal to $0.02735 <
0.05 = \alpha$, hence the null hypothesis of equal policy effort in 2001 and 2008 must be
rejected in favor of the hypothesis of greater policy effort in 2008.

## 4.2.2    A permutation test for paired samples

Let us now examine a permutation solution to our problem, again assuming that $F_1$ and $F_2$ are unknown and that the nonparametric families of distributions, to which $F_1$ and $F_2$ belong, contain only nondegenerate distributions including: discrete, continuous and mixed (Pesarin an Salmaso, 2010). Let us first observe that the null hypothesis $H_0 : Z_1 \overset{d}{=} Z_2$ implies that the two variables $Z_1$ and $Z_2$ are exchangeable within each unit with respect to the two occasions or to the two paired samples. In other words, in $H_0$ the two observed values of each unit are considered as if they were randomly assigned with the same probability to occasion (or sample) 1 and 2. This implies that the sign of each difference $D_i = X_{2i} - X_{1i}$, $i = 1, \ldots, n$, is considered as if it were randomly assigned and the probabilities of $+$ and $-$ are both equal to $1/2$. Thus a possible solution to the testing problem is to consider the statistic $T^* = \sum_{i=1}^{n} D_i^*$. Its permutation distribution conditional to the observed dataset, under assumption that $H_0$ is true, can be obtained by considering all the $2^n$ possible random attributions of plus or minus signs to the differences. According to the permutation distribution, the $p$-value can be computed considering that:

- when the alternative is $H_1 : Z_2 \overset{d}{<} Z_1$ the null hypothesis is rejected for small values of $T$;

- when the alternative is $H_1 : Z_2 \overset{d}{>} Z_1$ the null hypothesis is rejected for large values of $T$;

- when the alternative is $H_1 : Z_2 \overset{d}{\neq} Z_1$ the null hypothesis is rejected for large values of $|T|$.

It is worth noting that this permutation solution does not require homoscedasticity of differences among units. Furthermore it relaxes the assumption of continuity and it can also be applied in the presence of discrete variables.

Let us apply the permutation solution to the problem of the treatment for body weight loss where the weights of $n = 12$ subjects before and after the treatment were considered to test whether the treatment is effective at the significance level $\alpha = 0.05$ (Table 4.1). We can use the function `perm.sign(x,y,paired,alternative,B,fun)` also for the one-sample testing problem on symmetry. The necessary $R$ code is included in the files `"t2p.r"` and `"permsign.r"`. As a matter of fact the location problem for paired samples consists of testing for the symmetry of the variable difference $D$ around zero. The input data for the case of paired samples must be the arrays that include the sample data x and y, the specification of the type of test `paired=TRUE` (the default value is `FALSE`, necessary in the case of test on symmetry), the type of alternative (`"greater"`, `"two.sided"`, `"less"` or, to perform the test on symmetry, the default value `FALSE`), the number $B$ of resamplings (permutations) and the type of combining function for multiple tests (the default value is `FALSE`, suitable for the case of a single test such as in the case of two paired samples). The $R$ commands are:

```
> source("t2p.r")
> source("permsign.r")
> w_pre=c(117.48,83.90,92.29,103.89,94.94,121.87,
+ 105.71,87.07,100.00,91.30,98.99,96.78)
> w_post=c(90.20,87.46,87.15,76.16,89.19,93.94,
+ 94.69,81.57,98.39,99.07,103.49,100.98)
> x=array(w_pre,dim=c(length(w_pre),1))
> y=array(w_post,dim=c(length(w_pre),1))
> perm.sign(x,y,paired=TRUE,alternative="greater",B=10000)
##################################################################
# Permutation Test for paired samples
# $p.value
# 0.04279572
##################################################################
```

The resulting $p$-value is equal to 0.043 and leads to the rejection of the null hypothesis of no effect of the treatment at the significance level $\alpha = 0.10$ in favor of the alternative that weight after treatment is less than weight before treatment.

Let us consider the experiment to evaluate whether monitoring the heartbeat adversely affects the recovery of a runner, that is it induces a lower reduction in heart rate after 1 min from the end of a race (Table 4.2). The goal of the analysis is to test whether HRR is greater for athletes who do not monitor their heartbeat at the significance level $\alpha = 0.10$. Considering that we are in the presence of matched samples, the two-paired sample permutation test is a suitable solution. The $R$ commands for this problem are:

```
> source("t2p.r")
> source("permsign.r")
> HRR_no=c(33,43,50,48,36,31,41,45)
> HRR_yes=c(36,41,45,47,35,37,32,49)
> x=array(HRR_no,dim=c(length(HRR_no),1))
> y=array(HRR_yes,dim=c(length(HRR_yes),1))
> perm.sign(x,y,paired=TRUE,alternative="greater",B=10000)
##################################################################
# Permutation Test for paired samples
# $p.value
# 0.3913609
##################################################################
```

In this example the $p$-value is equal to $0.391 > 0.10 = \alpha$, thus the null hypothesis of no effect of monitoring the heartbeat on the HRR of runners cannot be rejected.

Regarding the problem where the industrial policy effort in prefectures of Guangdong (in the south of China) was measured through a suitable index and the goal of the study was to test whether the policy effort in 2008 is significantly greater than the policy effort in 2001 at the significance level $\alpha = 0.05$ (Table 4.3), the $R$ code for the permutation test is:

```
> source("t2p.r")
> source("permsign.r")
> china=read.csv("china.csv",header=TRUE,sep=";")
> Effort01=china[,2]
> Effort08=china[,3]
> x=array(Effort01,dim=c(length(Effort01),1))
> y=array(Effort08,dim=c(length(Effort08),1))
> perm.sign(x,y,paired=TRUE,alternative="less",B=10000)
################################################################
# Permutation Test for Symmetry
# $p.value
# 0.02219778
################################################################
```

In this example the $p$-value is equal to $0.022 < 0.05 = \alpha$, thus the null hypothesis of null difference between policy effort in 2001 and 2008 must be rejected in favor of the alternative hypothesis of greater policy effort in 2008.

## 4.3   Repeated measures tests

The test for repeated measures discussed in this section is a location problem which can be considered a generalization to $C$ samples of the two-sample problem for paired data presented in the previous section. Here we assume that the observations of the $C$ samples are realizations of $C$ non-independent variables. The data are $C$-dimensional vectors of replications and each vector represents the observed values of a $C$-dimensional multivariate variable. The goal of the study consists of testing the significance of a shift in location in some of the $C$ populations. The first assumption is that the subjects/objects are 'homogeneous' with respect to the most important experimental conditions, the so called covariates. The set of data vectors $(X_{1i}, \ldots, X_{Ci})$, $i = 1, \ldots, n$ may be viewed as a random sample of $n$ vectors from the multivariate random variable $Z = (Z_1, \ldots, Z_C)$ with $Z_1, \ldots, Z_C$ non-independent random variables with marginal CDFs $F_1, \ldots, F_C$, respectively. The hypotheses of the problem are

$$H_0 : \left\{ Z_1 \overset{d}{=} \cdots \overset{d}{=} Z_C \right\} \equiv \left\{ F_1(x) = \cdots = F_C(x) \; \forall x \in \mathcal{R} \right\}$$

against the alternative

$$H_1 : \left\{ Z_j \overset{d}{\neq} Z_k \right\} \equiv \left\{ F_j(x) \neq F_k(x) \text{ for some } j, k \text{ and for some } x \in \mathcal{R} \right\}.$$

## 4.3.1   Friedman rank test for repeated measures

A testing problem for repeated measures can be solved with solutions suitable also for unreplicated complete block designs, where data consist of $n$ blocks of vectors of $C$ observations, supposed to be realizations of $C$-variate continuous independent random variables. This is also the case of the nonparametric test based on ranks proposed by Friedman (1937).

When the multivariate distribution of $(Z_1, \dots, Z_C)$ is not known neither using asymptotic approximations nor when variances of different populations may not be equal, the Friedman test provides a nonparametric solution based on ranks (Kvam and Vidakovic, 2007).

Let $X_{ji}$ be the observation related to the $i$th unit and $j$th sample (treatment) and $R_{ji}$ be the rank of $X_{ji}$ in the set of observations related to the $i$th unit $\{X_{1i}, \dots, X_{Ci}\}$: formally $R_{ji} = \mathbb{R}_i(X_{ji})$. Let $R_{j\bullet}$ denote the mean rank for the $j$th treatment, that is $R_j = \sum_i R_{ji}$ and $R_{j\bullet} = R_j / n$. The Friedman test statistic is

$$
FR = \frac{12n}{C(C+1)} \sum_{j=1}^{C} \left( R_{j\bullet} - \frac{C+1}{2} \right)^2
$$

$$
= \left[ \frac{12}{nC(C+1)} \sum_{j=1}^{C} R_j^2 \right] - 3n(C+1).
$$

The null hypothesis must be rejected in favor of the alternative for large values of the test statistic, that is at the significance level $\alpha$ when $FR_{obs} \geq FR_\alpha$, where $FR_\alpha$ is such that the probability of rejection when $H_0$ is true does not exceed $\alpha$ and $FR_{obs}$ is the observed value of the test statistic. Alternatively the decision rule can be based on the $p$-value, by comparing it with $\alpha$ as usual. Under the null hypothesis the test statistic follows asymptotically a chi-square distribution with $C - 1$ degrees of freedom. In the case of ties a suitable correction to the test statistic must be applied.

Let us analyze an economic example related to the prices (in euros for a 1.5 l bottle) of $n = 14$ different types of soft drinks, sold in $C = 5$ different outlets, on a specific date, when neither promotion nor special discount was applied. We want to test if the price differs depending on the shop at the significance level $\alpha = 0.05$. The null hypothesis is $H_0 : \{Price_1 \overset{d}{=} \cdots \overset{d}{=} Price_5\}$ and the alternative hypothesis is $H_1 : \{H_0 \text{ is not true}\}$. The data are shown in Table 4.4.

To perform the test we can use the function `friedman.test(y)` that requires as input argument the matrix of the observations y. The $R$ code for the analysis and the output are:

```
> drink=read.csv("prices.csv",header=TRUE,sep=";")
> drink=drink[,-1]
> data=as.matrix(drink)
> friedman.test(data)
################################################################
# Friedman rank sum test
```

Table 4.4    Prices (in euros) of 14 soft drinks in 5 different outlets.

| Drinks | Outlet 1 | Outlet 2 | Outlet 3 | Outlet 4 | Outlet 5 |
|--------|----------|----------|----------|----------|----------|
| 1  | 6.97 | 5.58 | 4.37 | 4.45 | 6.20 |
| 2  | 4.65 | 6.13 | 5.02 | 5.29 | 6.42 |
| 3  | 5.05 | 6.46 | 3.89 | 5.40 | 6.01 |
| 4  | 3.70 | 4.41 | 5.49 | 5.89 | 5.09 |
| 5  | 6.19 | 5.04 | 5.34 | 5.32 | 4.43 |
| 6  | 3.91 | 3.94 | 5.47 | 6.69 | 6.73 |
| 7  | 4.44 | 5.20 | 4.26 | 4.25 | 3.30 |
| 8  | 5.23 | 5.96 | 5.09 | 4.72 | 5.23 |
| 9  | 5.52 | 7.18 | 4.41 | 5.54 | 6.24 |
| 10 | 5.52 | 5.78 | 5.17 | 5.26 | 5.01 |
| 11 | 4.57 | 2.67 | 4.97 | 4.01 | 6.34 |
| 12 | 6.00 | 4.61 | 4.22 | 7.56 | 6.20 |
| 13 | 5.87 | 5.05 | 3.87 | 4.14 | 5.89 |
| 14 | 4.63 | 4.97 | 4.16 | 6.52 | 3.81 |

```
# data: as.matrix(drink)
# Friedman chi-squared =5.1183, df=4, p-value = 0.2754
###################################################################
```

Among the outputs, the function `friedman.test` returns the $p$-value. In this case it is equal to $0.275 > 0.05 = \alpha$, hence the null hypothesis of no difference between the discount prices cannot be rejected.

Another example relates to a medical problem and a method for rating, patient's nutritional status, with a subjective global assessment (SGA). Considered factors include weight change, appetite or anorexia and other similar symptoms. There were $n = 12$ patients affected by chronic heart failure treated with a drug based on iron. Patients were monitored with a SGA after 6, 12 and 18 months. We want to test whether SGA changes over time at the significance level $\alpha = 0.01$. The null hypothesis is $H_0 : \left\{ SGA_{6\,months} \overset{d}{=} SGA_{12\,months} \overset{d}{=} SGA_{18\,months} \right\}$ and the alternative hypothesis is $H_1 : \{H_0$ is not true$\}$. The data are shown in Table 4.5.

The $R$ commands to perform the test and the output are:

```
> SGA=read.csv("iron.csv",header=TRUE,sep=";")
> SGA=SGA[,-1]
> data=as.matrix(SGA)
> friedman.test(data)
###################################################################
# Friedman rank sum test
# data: as.matrix(iron)
# Friedman chi-squared =15.1667, df=2, p-value = 0.0005089
###################################################################
```

Table 4.5   Subjective global assessment (0–100 scale) of 12 patients affected by chronic heart failure treated with iron, 6, 12 and 18 months after treatment.

| | Time after treatment (months) | | |
|---|---|---|---|
| Patient | 6 | 12 | 18 |
| 1 | 75.25 | 74.87 | 77.29 |
| 2 | 75.08 | 77.13 | 80.40 |
| 3 | 74.12 | 77.68 | 80.88 |
| 4 | 65.82 | 70.48 | 67.95 |
| 5 | 58.02 | 57.92 | 59.73 |
| 6 | 65.30 | 72.26 | 78.70 |
| 7 | 81.31 | 83.93 | 89.51 |
| 8 | 66.18 | 73.43 | 75.83 |
| 9 | 73.44 | 77.23 | 78.88 |
| 10 | 61.45 | 67.05 | 66.10 |
| 11 | 68.17 | 76.79 | 74.72 |
| 12 | 60.10 | 66.94 | 70.66 |

The $p$-value (0.001) is less than the significance level $\alpha$ (0.01), thus the null hypothesis of no change of the SGA over time must be rejected.

Let us consider now an example of teaching evaluations of $C = 3$ university courses (Economics degree) evaluated by $n = 20$ students with a rate from 0 to 100. We want to test whether one or more courses present significantly different evaluations at the significance level $\alpha = 0.05$. The null hypothesis is $H_0 : \{Rate_{Stat} \overset{d}{=} Rate_{Mat} \overset{d}{=} Rate_{Econ}\}$ and the alternative hypothesis is $H_1 : \{H_0$ is not true$\}$. The data are shown in Table 4.6.

As for the previous example, the $R$ code for the analysis is:

```
> rate=read.csv("students.csv",header=TRUE,sep=";")
> rate=rate[,-1]
> data=as.matrix(rate)
> friedman.test(data)
##########################################################################
# Friedman rank sum test
# data: as.matrix(student)
# Friedman chi-squared =17.1, df=2, p-value = 0.0001935
##########################################################################
```

According to the very small $p$-value the null hypothesis must be rejected in favor of the alternative hypothesis of differences in the evaluations of the three courses.

Table 4.6   Teaching evaluations of three university courses by students studying for an Economics degree.

| Student | Statistics | Mathematics | Econometrics |
|---|---|---|---|
| 1 | 88 | 50 | 81 |
| 2 | 36 | 26 | 76 |
| 3 | 90 | 59 | 84 |
| 4 | 86 | 48 | 80 |
| 5 | 94 | 98 | 99 |
| 6 | 70 | 29 | 96 |
| 7 | 69 | 49 | 66 |
| 8 | 74 | 62 | 45 |
| 9 | 39 | 42 | 91 |
| 10 | 71 | 38 | 85 |
| 11 | 56 | 40 | 63 |
| 12 | 68 | 65 | 95 |
| 13 | 92 | 73 | 52 |
| 14 | 79 | 44 | 75 |
| 15 | 100 | 67 | 97 |
| 16 | 93 | 82 | 54 |
| 17 | 87 | 43 | 77 |
| 18 | 58 | 41 | 60 |
| 19 | 89 | 46 | 64 |
| 20 | 72 | 30 | 83 |

## 4.3.2   A permutation test for repeated measures

A permutation testing solution in the presence of repeated measures is provided by the permutation test for related samples that can also be used for unreplicated complete block designs. No interaction is assumed between units and occasions of measurement (Pesarin and Salmaso, 2010) and the $n$ units are each observed $C$ times. Thus the data $\{X_{ji}; i = 1, \ldots, n; j = 1, \ldots, C\}$ are realizations of a nondegenerate response variable and the set of observations $\{X_{ji}; j = 1, \ldots, C\}$ is called the individual profile of the $i$th unit. Let us assume that:

1. the $n$ profiles are independent;

2. the within unit responses are homoscedastic and exchangeable with respect to treatments in the null hypothesis;

3. the underlying model for fixed effects is additive, that is $Z_{ji} = \theta + \beta_i + \mu_j + \epsilon_{ji}$, with $i = 1, \ldots, n$ and $j = 1, \ldots, C$, where $\theta$ is a population constant (e.g., unknown mean or median), $\beta_i$ is a unit specific constant (block effect corresponding to the $i$th unit), $\mu_j$ is the unknown effect of the $j$th treatment and

$\epsilon_{ji}$ are exchangeable random errors with zero mean or median with common distribution and standard deviation $\sigma_i$, which can vary with respect to units.

To test the null hypothesis of equal treatment effects $H_0 : [\mu_1 = \cdots = \mu_C]$ against the alternative $H_1 : [\mu_1, \ldots, \mu_C$ not all equal] we can consider that in $H_0$ data within each unit are exchangeable with respect to treatments, and that they are independent with respect to units. A suitable test statistic is:

$$T_{RS}^* = \frac{\sum\limits_{j=1}^{C} \left( \overline{X}_{j\bullet}^* - \overline{X}_{\bullet\bullet} \right)^2}{\sum\limits_{i=1}^{n} \sum\limits_{j=1}^{C} \left( X_{ji}^* - \overline{X}_{\bullet i} - \overline{X}_{j\bullet}^* + \overline{X}_{\bullet\bullet} \right)^2},$$

where $\overline{X}_{j\bullet}^* = \sum_i X_{ji}^*/n$, $j = 1, \ldots, C$, $\overline{X}_{\bullet i} = \sum_j X_{ji}^*/C$, $i = 1, \ldots, n$ and $\overline{X}_{\bullet\bullet} = \sum_{i,j} X_{ji}^*/(nC)$, respectively are column, row and global means with respect to the $n \times C$ data matrix after a given permutation. Since the permutations are performed within units with respect to the treatment levels, the unit means and the global mean are permutationally invariant, that is $\overline{X}_{\bullet i}^* = \overline{X}_{\bullet i}$ and $\overline{X}_{\bullet\bullet} = \overline{X}_{\bullet\bullet}^*$. The null hypothesis must be rejected for large values of $T_{RS}$, hence the $p$-value is the probability that $T_{RS}^*$ takes values greater than or equal to the observed value of $T_{RS}$.

This permutation test can be applied also when variances of responses are not equal among units (or blocks). It is also a valid solution in the presence of discrete variables, hence it is more flexible with respect to other nonparametric solutions. Furthermore through the nonparametric combination methodology, it provides a suitable and effective solution to the multivariate extension of this testing problem.

To solve the problem of prices (in euros for a 1.5 l bottle) of $n = 14$ different types of soft drinks, sold in $C = 5$ different outlets, on a specific date, when no promotion or special discount was applied (Table 4.4), where the goal is to test whether the price differs depending on the shop at the significance level $\alpha = 0.05$, we can apply the permutation test. We can use the function two.way.rs(data) suitable for a two-way analysis of variance design with related samples. The $R$ code and the result are:

```
> source("t2p.r")
> source("twoWayRs.r")
> drink=read.csv("prices.csv",header=TRUE,sep=";")
> drink=drink[,-1]
> two.way.rs(drink,B=10000)
####################################################################
# $pvalue
# 0.2434757
####################################################################
```

The $p$-value 0.243 is greater than $\alpha = 0.05$, hence the null hypothesis of no difference between the discount prices cannot be rejected.

Let us consider the medical problem related to SGA of nutritional status of patients affected by chronic heart failure and treated with a drug based on iron (Table 4.5). Patients were monitored with a SGA after 6, 12 and 18 months and the goal of the study is to test whether SGA changes over time at the significance level $\alpha = 0.01$.

The $R$ commands for the analysis and the result are:

```
> source("t2p.r")
> source("twoWayRs.r")
> SGA=read.csv("iron.csv",header=TRUE,sep=";")
> SGA=SGA[,-1]
> two.way.rs(SGA,B=10000)
###################################################################
# $pvalue
# 9.999e-05
###################################################################
```

The $p$-value is very significant and less than $\alpha = 0.01$, leading to rejection of the null hypothesis of no effect of the therapy over time. Finally, let us apply the permutation solution to the example of teaching evaluations of $C = 3$ university courses (Economics degree) evaluated by $n = 20$ students with a rate from 0 to 100, where the goal is to test whether one or more courses present different evaluations at the significance level $\alpha = 0.05$. The syntax in $R$ and the output are:

```
> source("t2p.r")
> source("twoWayRs.r")
> rate=read.csv("students.csv",header=TRUE,sep=";")
> rate=rate[,-1]
> two.way.rs(rate,B=10000)
###################################################################
# $pvalue
# 9.999e-05
###################################################################
```

Even in this case the $p$-value is very small and leads to rejection of the null hypothesis of no difference between the courses' rates and to conclude that courses present significant different evaluations.

# References

Arboretti Giancristofaro, R., Bonnini, S. and Salmaso, L. (2007) A performance indicator for multivariate data. Quaderni di Statistica, 9, 1–29.

Barbieri, E., Di Tommaso, M. and Bonnini, S. (2012) Industrial development policies and performances in southern China: beyond the specialised industrial cluster program. China Economic Review, 23, 613–625.

Friedman, M. (1937) The use of ranks to avoid the assumption of normality implicit in the analysis of variance. Journal of the American Statistical Association, 32, 675–701.

Hollander, M. and Wolfe, D.A. (1999) Nonparametric Statistical Methods, 2nd edn. John Wiley & Sons, Ltd.

Kvam, P.H. and Vidakovic, B. (2007) Nonparametric Statistics with Applications to Science and Engineering. John Wiley & Sons, Ltd.

Lago, A. and Pesarin, F. (2000) Nonparametric combination of dependent rankings with application to the quality assessment of industrial products. Metron, LVIII, 39-52.

Pesarin, F. (2001) Multivariate Permutation Tests with Applications in Biostatistics. John Wiley & Sons, Ltd.

Pesarin, F. and Salmaso, L. (2010) Permutation Tests For Complex Data: Theory, Applications and Software. John Wiley & Sons, Ltd.

Wilcoxon, F. (1945) Individual comparisons by ranking method. Biometrics, 1, 80-83.

# 5

# Tests for categorical data

## 5.1 Introduction

In real applications several phenomena are represented by categorical variables. When data are not numerical some specific testing problems may arise and the categorical nature of data implies the application of specific testing procedures.

In addition to the classical classification of problems according to the number of compared samples and the presence of one or more dependent response variables, we can also distinguish between problems for binary or nonbinary variables, depending on the number of observable categories (two or more). Hence the data of the problems presented in this chapter can be represented in the following ways:

- $\{X_i; i = 1, \dots, n\}$ (univariate one-sample problem where $n$ is the sample size);

- $\{X_{ji}; i = 1, \dots, n_j; j = 1, 2\}$ (univariate two-sample problem where $n_j$ is the size of the $j$th sample);

- $\{X_{ih}; i = 1, \dots, n; h = 1, \dots, q\}$ ($q$-variate one-sample problem where $n$ is the sample size).

The support of the variables is always represented by a set of two or more categories. When the categories are ordered (e.g. categorical judgments, educational level, age groups, etc.) the responses are said to be ordered categorical; otherwise they are said to be nominal categorical.

In multisample and/or bivariate problems, categorical data can be presented through contingency tables because the sample observations can be classified according to two factors. The levels of one factor correspond to the rows of the table; the levels of the other factor to the columns. The integer number $f_{jk}$, on the $j$th row and $k$th column of the observed table, represents the absolute joint frequency of factor 1

*Nonparametric Hypothesis Testing: Rank and Permutation Methods with Applications in R,*
First Edition. Stefano Bonnini, Livio Corain, Marco Marozzi and Luigi Salmaso.
© 2014 John Wiley & Sons, Ltd. Published 2014 by John Wiley & Sons, Ltd.
Companion website: http://www.wiley.com/go/hypothesis_testing

at level $j$ and factor 2 at level $k$. Sometimes the problems are defined according to the sizes of the contingency table.

A $R \times C$ contingency table, with $R, C \geq 2$, can be used to represent one-sample bivariate problems where the two factors take the role of component variables with $R$ and $C$ categories, respectively. In the presence of $R$ samples and a categorical variable with $C$ categories, a $R \times C$ contingency table can be used to represent data, if one factor takes the role of symbolic treatment denoting the sample and the other factor represents the response variable.

In Section 5.2 some one-sample tests on proportions are presented and the procedures of the binomial test and McNemar test (univariate and multivariate) are described. Section 5.3 is dedicated to problems for $2 \times 2$ contingency tables and specifically to the Fisher exact test and the permutation test for two proportions. Section 5.4 deals with more general problems related to $R \times C$ contingency tables with $R, C \geq 2$. In this final section, the Anderson–Darling permutation test, the permutation test on moments for ordinal variables, and the chi-square permutation test are discussed.

## 5.2   One-sample tests

A typical one-sample problem for binary variables, when only two categories are observed on the statistical units, consists of testing whether the proportion of units with a given characteristic in the population is equal to a specified value or not. For this problem, a binomial test on one proportion is a suitable procedure.

A bivariate extension of the previous problem characterizes the test for marginal homogeneity in the presence of two dependent binary responses. In this problem, the null hypothesis consists of the equality of marginal distributions of the two component variables. This type of problem with bivariate binary variables also covers the cases of repeated measures or dependent samples. A nonparametric solution is the McNemar test. A multivariate extension of this test for repeated measures or dependent samples is also given.

This section is dedicated to one-sample tests for dichotomous variables. The binomial test on one proportion, the McNemar test and its multivariate extension are presented.

### 5.2.1   Binomial test on one proportion

In several real problems, the data of interest are the proportions or percentages of statistical units that meet a certain characteristic: the proportion of defective production units in a statistical quality control of a production process, the percentage of shares in loss of a given financial package, the percentage of patients who had a disease remission, etc. Such data can be analyzed with a simple testing procedure based on the binomial distribution.

Let us consider a sample of $n$ data from a dichotomous categorical variable and suppose this variable can take two categories, conventionally defined as success and failure. The data can be also considered as the outcomes of $n$ independent symbolic trials. Let us assume independence among the trials and that in each trial the probability of success is equal to $\pi$ and that of failure is equal to $1 - \pi$. Formally we can represent data as follows:

$$X_i = \begin{cases} 1 & \text{if the } i\text{th observation is a success} \\ 0 & \text{if the } i\text{th observation is a failure,} \end{cases}$$

with $i = 1, \ldots, n$. When we are interested in testing wether $\pi$ is equal to a specific value $\pi_0$ against the alternative that it is different, a two-tailed binomial test can be applied where $H_0 : \pi = \pi_0$ and $H_1 : \pi \neq \pi_0$. A suitable test statistic for such a problem might be the number of successes $T = \sum_{i=1}^{n} X_i$. The null distribution of $T$ is $Bn(n, \pi_0)$ and the $p$-value is

$$\Pr\{|T| \geq |T_{obs}|\}.$$

Instead of the binomial distribution, the normal approximation (and a correction for continuity) can be used when $n \geq 20$ (Conover, 1999). When the hypotheses are one-sided, for example $H_0 : \pi \geq \pi_0$ and $H_1 : \pi < \pi_0$, then the $p$-value is given by $\Pr\{T \leq T_{obs}\}$. For the opposite case (upper-tailed test) we have a similar $p$-value but with the opposite inequality.

Even if alternative powerful procedures were proposed for testing this problem, the binomial test is often a preferable solution because there is a good compromise between simplicity and power (Conover, 1999). For estimation problems, confidence intervals for $\pi$ can be used (Clopper and Pearson, 1934). Alternative proposals are those of Anderson and Burstein (1967, 1968). For multinomial proportions Quesenberry and Hurst (1964) and Goodman (1965) propose simultaneous confidence intervals.

For example, let us consider a random sample of $n = 300$ pieces of finished product drawn from the production by the quality assurance office of a company that produces electronic components for computers. Some imperfections were found in 11 pieces of the sample. The company is interested in testing whether the percentage of rejects, that is the percentage of faulty pieces of the whole production, is equal to 3% against the alternative that it is greater than 3%. Thus the null hypothesis is $H_0 : \pi = 0.03$ and the alternative is $H_1 : \pi > 0.03$. The sample relative frequency of pieces that should be rejected is $T/n = 11/300 = 0.0367$.

The basic package of $R$ provides the function `binom.test(x,n,p, alternative)` that performs an exact test of a simple null hypothesis about the probability of success in a Bernoulli experiment. In particular x is the number of successes, or a vector of length 2 giving the numbers of successes and failures, respectively. The argument n is the number of trials (ignored if x has length 2). Parameter p is the probability of success hypothesized in $H_0$ (by default equal to 0.5). The final argument `alternative` denotes the alternative hypothesis and must be `"two.sided"`, `"greater"` or `"less"`. The $R$ code to perform the analysis is:

```
> binom.test(x=11,n=300,p=0.03,alternative="greater")
```

The final output is:

```
#####################################################################
# Exact binomial test
#
# data:  11 and 300
# number of successes = 11, number of trials = 300, p-value = 0.2922
# alternative hypothesis:
# true probability of success is greater than 0.03
# 95 percent confidence interval:
# 0.02069653 1.00000000
# sample estimates:
# probability of success
#         0.03666667
#####################################################################
```

The function returns a summary with the number of successes, the number of trials and the $p$-value of the test. The function returns also a confidence interval for the probability of success (95 percent confidence interval) and the estimated probability of success corresponding to the sample relative frequencies of successes. For the described problem the $p$-value is equal to 0.292 and the null hypothesis that the true percentage of damaged production units is equal to 3% should not be rejected.

Another interesting example concerns a financial problem. A bank wishes to test the profitability of a financial investment consisting of a package of shares. To control the risk associated with such type of investment, a random sample of $n = 20$ shares were drawn from the set of shares that had selectable from the package, the variations of the values after a specified time period were observed and a test on the percentage of shares that had fallen in value after the period was performed, to test whether this percentage is less than 30% (alternative hypothesis) at the significance level $\alpha = 0.05$. Consider that the binomial distribution requires independence of the $n$ sample observations, hence in this example we assume that, in the considered package, the performance of each share is independent from that of any other. The statistical hypotheses are $H_0 : \pi \geq 0.3$ against $H_1 : \pi < 0.3$. In this experiment only 2 of the 20 shares depreciated (observed relative frequency 0.10). The execution of the test with $R$ and the results are as follows:

```
> binom.test(x=2,n=20,p=0.3,alternative="less")
#####################################################################
# Exact binomial test
#
# data:  2 and 20
# number of successes = 2, number of trials = 20, p-value = 0.03548
# alternative hypothesis:
#           true probability of success is less than 0.3
```

```
# 95 percent confidence interval:
# 0.0000000 0.2826185
# sample estimates:
# probability of success
#           0.1
##########################################################################
```

The resulting $p$-value ($0.035 < \alpha = 0.05$) leads to reject the null hypothesis in favor of the alternative that the percentage of shares that will devalue is less than 30%.

Let us consider a medical example. In an observational study, $n = 86$ patients who have been diagnosed 'Dermatitis Herpetiformis', $T = 10$ have had, after a certain period of time, spontaneous disease remission (Paek $et\ al.$, 2011). We wish to test the hypothesis that more than 10% of patients recover spontaneously (alternative hypothesis) at the significance level $\alpha = 0.10$. The statistical hypotheses are $H_0 : \pi \leq 0.10$ against $H_1 : \pi > 0.10$. The $R$ code and the results are as follows:

```
>binom.test(x=10,n=86,p=0.1,alternative="greater")
##########################################################################
# Exact binomial test
#
# data:   10 and 86
# number of successes = 10, number of trials = 86,
#                                    p-value = 0.3571
# alternative hypothesis:
#          true probability of success is greater than 0.1
# 95 percent confidence interval:
# 0.06445987 1.00000000
# sample estimates:
# probability of success
#           0.1162791
##########################################################################
```

According to the resulting $p$-value (0.357), we cannot reject the null hypothesis that the percentage of patients who recover spontaneously is less than or equal to 10%.

## 5.2.2    The McNemar test for paired data (or bivariate responses) with binary variables

A typical problem, related to the bivariate extension of the problem described in the previous section, is the $test\ for\ marginal\ homogeneity$. Let us assume that the dataset consists of $n$ independent couples of data $\{X_{i1}, X_{i2}\}$, $i = 1, \dots, n$ and that the two responses can take only two categories, conventionally denoted by 0 and 1. An application could be framed in the case where $\{X_{i1}, X_{i2}\}$ denotes the presence/absence of two characteristics on the $i$th statistical unit. Another common application is that

Table 5.1 Probability distribution of the bivariate Bernoulli random variable.

|  |  | $X_1$ | | |
|---|---|---|---|---|
|  |  | 0 | 1 | |
| $X_2$ | 0 | $\pi_{00}$ | $\pi_{01}$ | $\pi_{0\bullet}$ |
|  | 1 | $\pi_{10}$ | $\pi_{11}$ | $\pi_{1\bullet}$ |
|  |  | $\pi_{\bullet 0}$ | $\pi_{\bullet 1}$ | 1 |

of classifications expressed according to a dichotomous scale by two evaluators on $n$ objects/subjects or $n$ items. Marginal homogeneity consists of equality of the marginal distributions of the bivariate response or the agreement between the two evaluators. In both problems, we assume that data are realizations of a bivariate Bernoulli random variable. By denoting with $\pi_{rs}$ the probability of observing the couple $(r, s)$, with $r, s \in \{0, 1\}$, the joint probability distribution can be represented by Table 5.1.

The testing problem can be formally represented by the hypotheses $H_0 : \pi_{\bullet 1} = \pi_{1\bullet}$ against $H_1 : \pi_{\bullet 1} \neq \pi_{1\bullet}$.

By denoting with $f_{rs}$ the absolute frequency of the couple $(r, s)$ in the observed sample, with $r, s \in \{0, 1\}$, the joint frequency distribution can be represented by Table 5.2. Note that this table is not properly a contingency table, hence the techniques described in the following sections cannot be applied. The more $f_{00} + f_{01}$ tends to be similar to $f_{00} + f_{10}$, in other words the difference between $f_{01}$ and $f_{10}$ is near to zero, the greater is the empirical evidence in favor of the hypothesis of marginal homogeneity and vice versa. Hence a suitable test statistic for such a test might be based on $(f_{01} - f_{10})$. For small sample sizes the test statistic might be

$$T = f_{01}.$$

If the null hypothesis of marginal homogeneity is true, $T$ follows a binomial distribution with parameters $f_{01} + f_{10}$ and 0.5, formally $T \sim Bn(f_{01} + f_{10}, 0.5)$. The null

Table 5.2 Absolute frequency distribution of a bivariate binary response variable.

|  |  | $X_1$ | | |
|---|---|---|---|---|
|  |  | 0 | 1 | |
| $X_2$ | 0 | $f_{00}$ | $f_{01}$ | $f_{00} + f_{01}$ |
|  | 1 | $f_{10}$ | $f_{11}$ | $f_{10} + f_{11}$ |
|  |  | $f_{00} + f_{10}$ | $f_{01} + f_{11}$ | $n$ |

Table 5.3   Absolute frequency distribution of customers of the Gammatel company interested in the 'world breaking news' service, before and after the trial period.

| Before | After | |
|--------|-------|------|
|        | Yes   | No   |
| Yes    | 3     | 2    |
| No     | 4     | 11   |

hypothesis must be rejected for large or small values of $T$. When $f_{01} + f_{10} > 20$ then

$$T = (f_{01} - f_{10})^2 / (f_{01} + f_{10})$$

is usually considered as the test statistic (Kvam and Vidakovic, 2007). Under $H_0$ it follows a $\chi^2$ distribution with 1 degree of freedom. In some works the following version with discontinuity correction is proposed: $T = (|f_{01} - f_{10}| - 1)^2 / (f_{01} + f_{10})$ but, from a practical point of view, by using the computational capabilities of modern computers, it is not necessary. Trivial adaptations to the decision rule can be applied for the one-sided tests. This test was proposed by McNemar (1947). Some variations are proposed by Bennett and Underwood (1970), Mantel and Fleiss (1975), and McKinlay (1975) and Ury (1975). The McNemar test can be considered the extension of the one-sample test on proportion to the case of two dependent samples. It can be also considered a special case of the sign test for paired data.

For example, the telephone company Gammatel carried out a survey where a sample of customers was asked whether they were interested or not in the paid service 'world breaking news' on their phone. After a trial period in which the service was offered for free, the same customers were asked again about their interest in the service. The company wished to know whether the proportion $\pi$ of customers interested in the service was different after the trial period. Data, in the form of a $2 \times 2$ table, are shown in Table 5.3.

The command mcnemar.test(x) performs the McNemar test for symmetry of rows and columns for $2 \times 2$ tables. The R code for the analysis and the output are as follows:

```
> mobile=matrix(c(3,4,2,11),nrow=2,dimnames=list("Before"=
+ c("Yes","No"),"After"=c("Yes","No")))
> mcnemar.test(mobile)
################################################################
# McNemar's Chi-squared test with continuity correction
# data: mobile
# McNemar's chi-squared = 0.1667, df = 1, p-value = 0.6831
################################################################
```

The function returns the value of McNemar's statistic, the degrees of freedom of the approximate chi-square distribution of the test statistic (df) and the $p$-value of the test. By default, the function applies the continuity correction when computing the test statistic. If this correction is not wanted, the argument correct=FALSE should be specified. The resulting $p$-value 0.683 leads to not rejecting the null hypothesis of no effect of the trial period.

### 5.2.3   Multivariate extension of the McNemar test

This section is dedicated to a multivariate extension of the testing solution illustrated in Section 5.2.2. Let us consider the case of multivariate paired data with $q$ binary variables. The dataset is $\{(X_{1ih}, X_{2ih}), i = 1, \ldots, n; h = 1, \ldots, q\}$ and $\pi_{rs,h}$ denotes the probability/proportion of the couple $(r, s)$ for the $h$th variable, with $r, s \in \{0, 1\}$ and $h = 1, \ldots, q$. The multivariate testing problem can be defined as

$$H_0 : \bigcap_{h=1}^{q} [\pi_{01,h} = \pi_{10,h}]$$

against

$$H_1 : \bigcup_{h=1}^{q} [\pi_{01,h} <\neq> \pi_{10,h}],$$

where in the global alternative some of the partial hypotheses can be two-sided and others one-sided. Each partial hypothesis can be solved with the binomial test based on the test statistic $T_h = f_{01,h}$ which under $H_0$ follows a binomial distribution with parameters $f_{01,h} + f_{10,h}$ and 0.5, where $f_{rs,h}$ denotes the sample absolute frequency of the couple $(r, s)$ for the $h$th variable, with $r, s \in \{0, 1\}$ and $h = 1, \ldots, q$.

Equivalently, we can consider the following data transformation

$$Y_{ih} = g\left(X_{1i,h}, X_{2i,h}\right) = \begin{cases} +1 & \text{if} \quad X_{1i,h} < X_{2i,h} \\ -1 & \text{if} \quad X_{1i,h} > X_{2i,h} \\ 0 & \text{otherwise,} \end{cases}$$

and apply the permutation test for paired data based on the test statistic

$$T_h^* = \sum_{i=1}^{n} Y_{ih} S_i^*$$

with $S_i^* = +1$ with probability 0.5 and $-1$ with probability 0.5 under $H_0$. The application of the nonparametric combination (NPC) methodology for multivariate permutation tests allows this testing problem to be solved by performing $B$ permutations (randomly generating $B$ sets of $n$-dimensional vectors of signs), combining the $q$ significance level functions of the partial tests with a suitable combining function and computing the permutation $p$-value of the global test related to the combined test statistic. Hence this procedure can be considered a particular case of the more general problem concerning multivariate paired observations.

All partial tests are marginally unbiased, that is, for each of them, the probability of rejecting the null hypothesis in favor of the alternative when the specific alternative is true is greater than the significance level $\alpha$, because each of them is separately related to one component variable, and the NPC method provides a proper overall solution (Pesarin and Salmaso, 2010). Even if each partial test is binomially distributed, the multivariate (global) test is not multinomial. For this test we cannot use asymptotic approximations unless we know the dependence relations among component binomials (see Pesarin, 2001, for a deeper discussion), thus a conditional Monte Carlo approach based on $B$ iterations seems to be a suitable solution.

Let us take into account the same example considered in the previous section but with a multivariate extension. Assume the telephone company Gammatel interviewed a sample of 20 customers to know whether they were interested or not in both of the paid services 'world breaking news' (WBN) and 'national sport news' (NSN). After a free trial period, the sample was interviewed again asking the same question. Data are shown in Table 5.4. We want to test if the proportion of customers interested after the trial period is different from the proportion before the trial period.

We are in the presence of paired samples with binary variables ('0' for 'no' and '1' for 'yes'). In fact the variable is the same, observed on two different occasions. Thus we can consider the permsign function illustrated for the test on symmetry. The $R$ code for the analysis and the output are as follows:

```
> source("t2p.r")
> source("permsign.r")
> source("comb.r")
> data=read.csv("gammatel.csv",header=TRUE,sep=";")
> data.01=abs(data=="yes")
> wbn.diff=data.01[,2]-data.01[,1]
> nsn.diff=data.01[,4]-data.01[,3]
> x=array(c(wbn.diff,nsn.diff),dim=c(20,2))
> perm.sign(x,B=10000,fun="F")
###################################################################
# Multivariate Permutation Test
# Combination Function: Fisher
# $p.value
# [1] 0.1823818
###################################################################
```

The rough data displayed in Table 5.4 are included in the file gammatel.csv as yes/no answers of the respondents: in the $20 \times 4$ matrix each row corresponds to a respondent and for each respondent the interest (yes or no) for the WBN service before and after the free trial period and the interest (yes or not) before and after the free trial period for the NSN service are reported. The original (binary) categorical variables (in data) are then transformed into dichotomous variables (yes=1; no=0) stored in

Table 5.4   Absolute frequency distribution of customers of the Gammatel company interested in the 'world breaking news' (WBN) and 'national sport news' (NSN) services, before and after the trial period.

| WBN | | NSN | |
|---|---|---|---|
| Before | After | Before | After |
| No | No | No | No |
| No | No | No | No |
| No | No | No | No |
| No | No | No | No |
| No | No | No | No |
| No | No | No | No |
| No | No | No | No |
| No | Yes | No | No |
| No | Yes | No | No |
| No | Yes | No | No |
| No | Yes | No | Yes |
| Yes | No | No | No |
| Yes | No | Yes | No |
| Yes | Yes | No | No |
| Yes | Yes | No | No |
| No | No | Yes | No |
| No | No | Yes | No |
| No | No | Yes | No |
| No | No | Yes | No |
| Yes | Yes | Yes | Yes |

data.01 and the after–before differences are computed for each component of the bivariate response, such that wbn.diff takes value 1 if a customer not interested in the WBN service before becomes interested after the trial period, it takes value −1 if a customer interested before is no longer interested after the trial period and value 0 if a customer does not change their opinion. The variable nsn.diff, concerning the NSN service, is defined similarly. The input for the perm.sign function, also useful for the test on symmetry, are the multivariate array x of the computed differences, the number B of permutations and the type of combining function ("F"=Fisher, "L"=Liptak and "T"=Tippett) to combine the univariate components of the multivariate response and obtain a global $p$-value for the test. In this case the Fisher combining function is used. According to the $p$-value 0.182, the null hypothesis of equality of the proportions of interested customers before and after the trial period cannot be rejected.

## 5.3    Two-sample tests on proportions or $2 \times 2$ contingency tables

A $2 \times 2$ contingency table is a matrix of integer numbers representing frequencies with two rows and two columns. Each statistical unit can be classified according to two categorical variables or factors and each factor can take two levels (categories). Each row of the table corresponds to a level of one factor and each column to a level of the other factor. Hence the number on the $j$th row and $k$th column of the table represents the joint absolute frequency $f_{jk}$ related to the $j$th level of the first factor and the $k$th level of the second factor, with $j, k = 1, 2$.

In the presence of two independent populations, and thus two independent samples, the real or symbolic treatment which identifies the two samples takes the role of one factor. In this case, the dataset is given by $\{X_{ji}; i = 1, \ldots, n_j; j = 1, 2\}$, where $j$ denotes the sample and $i$ denotes the statistical unit. Let us assume that the response variable can take the two categorical modalities $A_1$ or $A_2$. The observed contingency table is represented in Table 5.5.

A typical testing problem consists of comparing the proportions of observations equal to $A_1$ in the two populations, formally indicated as $\pi_{1/1}$ and $\pi_{1/2}$. This problem corresponds to the univariate McNemar test for independent populations.

In the presence of two response variables observed on one sample, the dataset is given by $\{X_{ih}; i = 1, \ldots, n; h = 1, 2\}$, where $i$ denotes the statistical unit and $h$ denotes the variable. In other words, we are in the presence of a bivariate response variable and each component variable is binary. If $X_{i1}$ and $X_{i2}$ can take values in $\{A_1, A_2\}$ and $\{B_1, B_2\}$, respectively, then the support of the bivariate response is given by $\{A_1, A_2\} \times \{B_1, B_2\}$ and the observed contingency table is described in Table 5.6.

In this case, a typical problem consists of testing for the independence of the two variables, that is the null hypothesis that the joint proportion/probability of observing $(A_j, B_k)$ is equal to the product of the marginal proportion/probability of observing $A_j$ and the marginal proportion/probability of observing $B_k$. Formally we can write $H_0 : \pi_{jk} = \pi_{j\bullet} \times \pi_{\bullet k}$ and $H_1 : \pi_{jk} \neq \pi_{j\bullet} \times \pi_{\bullet k}$.

Even if the two problems for categorical data are conceptually different, they are very similar because based on the common idea that each observation is classified according to two factors (one of them can be the sample from which it comes) and

Table 5.5    Contingency table in the presence of two independent samples and binary response variable.

|            | $A_1$    | $A_2$    |       |
| ---------- | -------- | -------- | ----- |
| Sample 1   | $f_{11}$ | $f_{12}$ | $n_1$ |
| Sample 2   | $f_{21}$ | $f_{22}$ | $n_2$ |
|            | $f_{\bullet 1}$ | $f_{\bullet 2}$ | $n$ |

Table 5.6   Contingency table in the
presence of two binary response variables.

|        | $A_1$    | $A_2$    |          |
|--------|----------|----------|----------|
| $B_1$  | $f_{11}$ | $f_{12}$ | $f_{1\bullet}$ |
| $B_2$  | $f_{21}$ | $f_{22}$ | $f_{2\bullet}$ |
|        | $f_{\bullet 1}$ | $f_{\bullet 2}$ | $n$ |

the marginal absolute frequencies of the table are fixed. As a matter of fact equality
of sample proportions in the former problem is equivalent to independence of the
latter. By denoting $n_j$ with $f_{j\bullet}$ we have

$$\frac{f_{11}}{n_1} = \frac{f_{21}}{n_2} \Leftrightarrow \frac{f_{jk}}{n_j} = \frac{f_{\bullet k}}{n} \Leftrightarrow f_{jk} = \frac{f_{j\bullet} f_{\bullet k}}{n} \Leftrightarrow \frac{f_{jk}}{n} = \left(\frac{f_{j\bullet}}{n}\right) \left(\frac{f_{\bullet k}}{n}\right).$$

Hence a testing solution useful for one problem can also be applied to the other.
One solution is represented by the chi-square test. This solution is presented in the
nonparametric permutation version in Section 5.4 dedicated to the general case of
$R \times C$ contingency tables with $R, C \geq 2$. Two other solutions are the Fisher exact test
and the permutation test for comparing two proportions.

## 5.3.1   The Fisher exact test

The null hypothesis of the Fisher exact test consists of the independence between
two binary variables (or two-level factors) conditional on the marginal frequencies
of the $2 \times 2$ observed contingency table. The classification of the $n$ observations
according to the two factors are summed up in Table 5.7, where $f$ denotes the number
of observations classified in the cell in the first row and first column. Conditional on
the total number of observations $n$ and the marginal observed frequencies of the first
row $f_{1\bullet}$ and first column $f_{\bullet 1}$, the other joint frequencies in the table depend only on $f$.

A suitable test statistic for such a problem is $T = f$. Under the null hypothesis of
equality of the proportion of observations in the first row classified in the first column

Table 5.7   Contingency table for Fisher's exact test.

|          | Factor 1 | | |
|----------|----------|----------|----------|
| Factor 2 | Level 1  | Level 2  |          |
| Level 1  | $f$             | $f_{1\bullet} - f$                        | $f_{1\bullet}$   |
| Level 2  | $f_{\bullet 1} - f$ | $n - f_{1\bullet} - f_{\bullet 1} + f$ | $n - f_{1\bullet}$ |
|          | $f_{\bullet 1}$    | $n - f_{\bullet 1}$                       | $n$              |

and proportion of observations in the second row classified in the first column, the distribution of $T$ is hypergeometric, hence for $t = 1, \ldots, \min\left(f_{1\bullet}, f_{\bullet 1}\right)$:

$$
\Pr\{T = t\} = \frac{\binom{f_{1\bullet}}{t}\binom{n-f_{1\bullet}}{f_{\bullet 1}-t}}{\binom{n}{f_{\bullet 1}}}.
$$

For the two-tailed test, the $p$-value is equal to twice the minimum between $\Pr\{T \leq T_{obs}\}$ and $\Pr\{T \geq T_{obs}\}$; for the directional test it is $\Pr\{T \leq T_{obs}\}$ or $\Pr\{T \geq T_{obs}\}$ depending on whether the test is lower tailed or upper tailed, where $T_{obs}$ is the observed value of the test statistic.

We discuss now an application. A metallurgical company performs deformation experiments on two types of bars ($A$ and $B$) for seismic reinforcement produced with alloys with shape memory. For each type of bar $n = 10$ trials are carried out on a sample of bars drawn from the whole production and for each trial the bar is classified as 'Non-defective' (0) or 'Defective' (1). The company wishes to test the hypothesis that in the whole production the proportion of type $A$ defective bars is greater than the proportion of type $B$ defective bars. By denoting with $\pi_A$ and $\pi_B$ the respective proportions, the hypotheses are: $H_0 : \pi_A = \pi_B$ and $H_0 : \pi_A > \pi_B$. The significance level is $\alpha = 0.05$. The observed contingency table is shown in Table 5.8. The basic package of $R$ provides the function fisher.test(x,alternative) that performs Fisher's exact test in a contingency table with fixed marginals, where x is a two-dimensional contingency table in matrix form and alternative indicates the alternative hypothesis and must be one of "two.sided", "greater" or "less". Let us consider the $R$ code for the problem being studied:

```
> x=matrix(c(7,3,2,8),ncol=2,byrow=TRUE,
+ dimnames=list("type"=c("A","B"),"defective"=c("yes","no")))
> x
defective
type yes no
A    7   3
B    2   8
> fisher.test(x,alternative="greater",conf.int=FALSE)
#######################################################################
# Fisher's Exact Test for Count Data
#
# data:  x
# p-value = 0.03489
# alternative hypothesis: true odds ratio is greater than 1
# sample estimates:
# odds ratio
#    8.153063
#######################################################################
```

Table 5.8   Contingency table of deformation experiment on two types of bars (*A* and *B*).

| Type of bar | Defective | Non-defective | |
|---|---|---|---|
| *A* | 7 | 3 | 10 |
| *B* | 2 | 8 | 10 |
| | 9 | 11 | 20 |

Table 5.9   Contingency table of dog food experiment on two breeds of dogs asked to choose between food with or without flavoring additive.

| Breed | Food with flavoring additive | | |
|---|---|---|---|
| | Yes | No | |
| Breed 1 | 8 | 4 | 12 |
| Breed 2 | 5 | 3 | 8 |
| | 13 | 7 | 20 |

The function returns the *p*-value of the test, a confidence interval for the odds ratio if argument `conf.int = TRUE` and an estimate of the odds ratio. Note that in this example the *p*-value 0.035 leads to the rejection of the null hypothesis that the production of type *A* and type *B* bars present the same proportion of defective pieces in favor of the alternative hypothesis that the proportion of defective bars is greater for type *A*.

A second application is related to a producer of dog food who performed an experiment on two samples of dogs of two different breeds requiring each animal to choose between a food containing a flavoring additive and another with the same ingredients but without additive. We wish to test if the proportion of dogs that chooses food with additive is equal or not in the two breeds at the significance level $\alpha = 0.05$. The observed contingency table is shown in Table 5.9. Similarly to the previous example the *R* code is:

```
> x=matrix(c(8,4,5,3),ncol=2,byrow=TRUE,
+ dimnames=list("Breed"=c("Breed 1","Breed 2"),
  "Additive"=c("yes","no")))
> x
          Additive
Breed    yes no
  Breed 1  8  4
  Breed 2  5  3
> fisher.test(x,alternative="two.sided",conf.int=FALSE)
########################################################################
```

```
# Fisher's Exact Test for Count Data
#
# data:   x
# p-value = 1
# alternative hypothesis: true odds ratio is not equal to 1
# sample estimates:
# odds ratio
#   1.189031
###########################################################################
```

In this example, according to the $p$-value, there is no empirical evidence that the proportion of dogs that choose the food with flavoring additive is different in the two breeds.

## 5.3.2    A permutation test for comparing two proportions

An alternative solution to the Fisher exact test is the permutation test for $2 \times 2$ contingency tables. This testing procedure is similar to the one related to the two-sample permutation test on central tendency. To see this similarity, consider the problem of comparing the proportions of two populations. Imagine factor 2 of Table 5.7 as the symbolic treatment useful for defining the two samples and factor 1 as the categorical (binary) variable taking values in $\{A_1, A_2\}$ (Table 5.5). Without loss of generality consider being interested in testing $H_0 : \pi_{1/1} = \pi_{1/2}$ against $H_1 : \pi_{1/1} <\neq> \pi_{1/2}$, where $\pi_{1/j}$ is the proportion/probability related to $A_1$ in population $j$ ($j = 1, 2$). A suitable test statistic might be $T = f_{11}/n_1 - f_{21}/n_2$, where $f_{j1}$ is the absolute frequency of $A_1$ in sample $j$, with the rejection region and rule for computing the $p$-value depending on $H_1$. Under the null hypothesis, exchangeability holds and we can estimate the permutation distribution of $T$ by performing $B$ random permutations, computing the corresponding values $T^*$ and then obtaining the permutation $p$-value as usual.

Consider now the following data transformation where the support of the original variable $\{A_1, A_2\}$ is replaced by the set $\{1, 0\}$:

$$Z_{ji} = \begin{cases} 1 & \text{if } X_{ji} = A_1 \\ 0 & \text{if } X_{ji} = A_2, \end{cases}$$

where $X_{ji}$ is the observation on the $i$th unit of the $j$th sample. Thus $f_{11}$ and $f_{21}$ denote the numbers of 1s observed in the first and second sample, respectively, with reference to variable $Z$. Hence the test statistic $T = f_{11}/n_1 - f_{21}/n_2$ can be considered as the difference of sample means of a binary variable $T = \overline{Z}_1 - \overline{Z}_2$ and the testing procedure is the same as for the two-sample permutation test on central tendency.

Let us apply this method to the deformation experiment on two types of bars shown in Table 5.8. For each type of bar $n = 10$ trials were carried out and for each trial the bar is classified as defective (1) or non-defective (0). We want to test whether the proportion of defective bars of type $A$ is greater at the significance level $\alpha = 0.05$.

Let us consider the same code used to compare two independent samples in terms of location, that is the function perm.2samples(data,alt,B). The *R* syntax is:

```
> source("t2p.r")
> source("perm_2samples.r")
> data.bin=array(c(rep(c(1,2),each=10),
+ 1,0,1,1,1,0,1,1,1,0,0,1,0,0,0,0,0,1,0,0),dim=c(20,2))
> T=perm.2samples(data.bin,alt="greater",B=10000)
> T$p.value
[1] 0.03379662
```

Note that we have defined an array of data (data.bin) in which the first column contains the labels of the groups (1=type *A* bar; 2=type *B* bar) and the second column contains the observed data for each single experimental bar, as required by the function perm.2samples. The other required input are the type of alternative alt ("greater", "less" or "two.tailed") and the number *B* of permutations.

The significance levels are obtained with the function t2p. In our case, the alternative is "greater" and 10 000 permutations are considered. As in the Fisher exact test, the resulting *p*-value (0.034) is less than $\alpha$ and the null hypothesis of no difference between the two groups should be rejected in favor of the alternative that the proportion of type *A* defective bars is greater.

Consider now the example of the producer of dog food and the experiment on two samples of dogs of different breeds which have been asked to choose between a food containing a flavoring additive and another with the same ingredients but without additive, see Table 5.9. We want to test whether the proportions of dogs of the two breeds which prefer food with flavoring additive are not equal at $\alpha = 0.05$. Similarly to the previous example the code is:

```
> source("t2p.r")
> source("perm_2samples.r")
> data.dog=array(c(rep(c(1,2),c(12,8)),
+ 1,0,1,1,0,1,0,1,1,1,1,0,0,1,0,1,1,0,1,1),dim=c(20,2))
> T=perm.2samples(data.dog,alt="two.sided",B=10000)
> T$p.value
[1] 1
```

According to the *p*-value, also using the permutation test, the null hypothesis of equality of the proportions cannot be rejected.

## 5.4    Tests for $R \times C$ contingency tables

In this section, the problems related to $2 \times 2$ contingency tables are extended to the general case of $R \times C$ contingency tables with $R, C \geq 2$. Also this case can be related to multisample problems for categorical variables or to a one-sample problem for a

Table 5.10  Contingency table with $R$ rows and $C$ columns.

| Factor 2 | Factor 1 | | | |
|---|---|---|---|---|
| | Level 1 | ... | Level $C$ | |
| Level 1 | $f_{11}$ | ... | $f_{1C}$ | $f_{1\bullet}$ |
| ... | ... | ... | ... | ... |
| Level $R$ | $f_{R1}$ | ... | $f_{RC}$ | $f_{R\bullet}$ |
| | $f_{\bullet 1}$ | ... | $f_{\bullet C}$ | $n$ |

bivariate categorical variable. In the former case, the two table sizes represent the number of samples and the number of categories of the response variable. In the latter case, they represent the number of categories of the first and of the second variable.

In other words, the contingency table can be represented by Table 5.10. As in the particular case of $2 \times 2$ contingency tables, in some problems factor 2 denotes the symbolic treatment representing the sample of origin of each observation (multisample problems) and factor 1 denotes the response variable. In the bivariate problems each factor denotes a component of the bivariate response.

The first type of problem considered in this section is the *goodness-of-fit* for ordered categorical variables, which includes two-sample and multisample problems with one-sided and two-sided alternatives. For these problems the Anderson–Darling permutation test and the permutation test on moments are described. Secondly, the test for independence and the solution of the chi-square permutation test are presented.

## 5.4.1   The Anderson–Darling permutation test for $R \times C$ contingency tables

The *goodness-of-fit* problem for categorical variables is very common in real problems and one of the oldest from the methodological point of view. Let us consider the two-sample problem for ordered categorical variables with a one-sided alternative hypothesis. In the specialized literature this type of alternative hypothesis sometimes takes the name of *stochastic dominance*. This is a complex problem with no easy solution especially among the parametric methods and in particular the likelihood ratio approach (Sampson and Whittaker, 1989; El Barmi and Dykstra, 1995; Wang, 1996; Cohen and Sackrowitz, 1998; El Barmi and Mukerjee, 2005). Within the maximum likelihood ratio solutions the asymptotic distribution of the test statistic under the null hypothesis depends on the true unknown parameters and this makes its practical application very difficult.

Let us assume that the support of the response variable is $\{A_1, \ldots, A_C\}$, with $A_k$ ($k = 1, \ldots, C$) ordered categories such that if $1 \leq r < s \leq C$ then $A_r \prec A_s$, that is, if the categories correspond to judgments, $A_s$ is a better judgment than $A_r$, if the categories represent classes of numerical values, $A_s$ corresponds to higher values than $A_r$, etc. The data of the problem are $\{X_{ji}; i = 1, \ldots, n_j; j = 1, 2\}$, where $X_{ji}$ denotes the $i$th observation in the $j$th sample.

Let us assume that sample data are realizations of two ordered categorical random variables taking values in $\{A_1, \ldots, A_C\}$, with cumulative distribution function (CDF) $F_1(x)$ and $F_2(x)$ for the first and second sample, respectively. The null hypothesis of the problem consists of the equality in distribution of the two random variables, that is the sample data come from the same population. For example, if the variables represent categorical judgments about satisfaction (e.g., very dissatisfied, moderately dissatisfied, etc.) the null hypothesis states that the satisfactions of the two groups are equal, that is the distributions of the judgments are equal. Formally

$$H_0 : F_1(A_k) = F_2(A_k), \ k = 1, \ldots, C.$$

Without loss of generality, let us assume an interest in testing whether in the first population the evaluations tend to be better. For nondirectional alternatives the chi-square test has to be used (Section 5.4.3). In the example related to satisfaction judgments we can say that the satisfaction of the first population tends to be higher than that of the second population. This alternative hypothesis of stochastic dominance can be formalized in terms of CDFs as

$$H_1 : F_1(A_k) \leq F_2(A_k), \ k = 1, \ldots, C \text{ and } F_1(A_k) < F_2(A_k) \text{ for at least one } k.$$

From the frequencies of the contingency table it is possible to compute the table of cumulative absolute frequencies shown in Table 5.11. The symbol $N_{jk}$ denotes the cumulative absolute frequency of the $k$th category for the $j$th sample, with $k = 1, \ldots, C$ and $j = 1, 2$. Formally $N_{jk} = \sum_{s=1}^{k} f_{js}$, from which follows that $N_{j1} = f_{j1}$ and $N_{jC} = n_j$ with $j = 1, 2$. Similarly we have $N_{\bullet k} = \sum_{s=1}^{k}(f_{1s} + f_{2s}) = \sum_{s=1}^{k} f_{\bullet s}$ and $N_{\bullet C} = n_1 + n_2 = n$. The sampling estimate of the CDF is $\widehat{F}_j(A_k) = N_{jk}/n_j, k = 1, \ldots, C, j = 1, 2$.

Under $H_0$, the data of the two samples are exchangeable, hence the permutation distribution of the test statistic can be obtained by permuting the rows of the dataset $X = \{X_{ji}; i = 1, \ldots, n; n_1, n_2\}$, that is considering all the possible tables with the same marginal frequencies $\{n_1, n_2, f_{\bullet 1}, \ldots, f_{\bullet C}\}$.

A suitable test statistic might be based on the differences $N_{2k} - N_{1k}$ and leads to the rejection of $H_0$ in favor of the stochastic dominance hypothesis for large values. For each permutation of the dataset, the test statistic based on the discrete

Table 5.11    Table of cumulative absolute frequencies in the presence of two independent samples and ordered categorical response variable.

|          | $A_1$    | ...  | $A_C$    |
|----------|----------|------|----------|
| Sample 1 | $N_{11}$ | ...  | $N_{1C}$ |
| Sample 2 | $N_{21}$ | ...  | $N_{2C}$ |
|          | $N_{\bullet 1}$ | ... | $n$      |

version of the Anderson–Darling goodness-of-fit test can be computed considering the corresponding permuted table:

$$T^* = \sum_{k=1}^{C-1} \left(N_{2k}^* - N_{1k}^*\right) \left[4\frac{N_{\bullet k}}{n} \left(\frac{n - N_{\bullet k}}{n}\right) \frac{n_1 n_2}{n-1}\right]^{-1/2},$$

where $N_{2k}^*$ and $N_{1k}^*$ denote the cumulative absolute frequencies for the $k$th category in the permuted table. The $p$-value of the test can be computed as the probability that the test statistic takes values greater than or equal to the observed one, according to the permutation distribution, as usual.

In the case of the two-sided test, that is when the alternative hypothesis states that the two distributions are not equal, we can call it the non-dominance alternative and formally we have

$$H_1 : F_1(A_k) \neq F_2(A_k), \text{ for at least one } k.$$

The suitable test statistic for this problem is

$$T^* = \sum_{k=1}^{C-1} \left(N_{2k}^* - N_{1k}^*\right)^2 \left[4\frac{N_{\bullet k}}{n} \left(\frac{n - N_{\bullet k}}{n}\right) \frac{n_1 n_2}{n-1}\right]^{-1}$$

and the null hypothesis is rejected for high values of the test statistic thus the $p$-value is computed as in the previous problem.

In the multisample case, when $R$ independent samples are compared (with $R > 2$) and the alternative hypothesis states that at least one population has a different distribution, the procedure is the same but the value of the test statistic for a given permutation of the dataset is

$$T^* = \sum_{j=1}^{R} \sum_{k=1}^{C-1} \left(\frac{N_{jk}^*}{n_j} - \frac{N_{\bullet k}}{n}\right)^2 \left[\frac{N_{\bullet k}}{n} \left(\frac{n - N_{\bullet k}}{n}\right) \frac{n - n_j}{n_j}\right]^{-1},$$

where $N_{jk}^*$ is the cumulative absolute frequency for the $k$th category in the $j$th sample in the permuted table with $j = 1, \ldots, R$, $n_j$ is the size of the $j$th sample (total of the $j$th row in the contingency table) and $N_{\bullet k} = \sum_{j=1}^{R} N_{jk} = \sum_{j=1}^{R} \sum_{s=1}^{k} f_{js} = \sum_{j=1}^{R} \sum_{s=1}^{k} f_{js}^*$.

Consider a market survey in which a sample of men and a sample of women were asked to say how much they liked drinking wine. The goal is to test whether the level of liking of males is higher than that of females (stochastic dominance alternative) at the significance level $\alpha = 0.01$. Note that the sensory evaluation was provided considering an ordinal scale with the seven categories: 'dislike extremely', 'dislike very much', 'dislike slightly', 'neutral', 'like slightly', 'like very much' and 'like extremely'. In the dataset the categories are represented by the numerical labels '1', '2', '3', '4', '5', '6', '7', respectively. The function ad.perm(data,B,alt) performs the two-sample Anderson–Darling permutation test just described. This function requires the following inputs: the matrix of data (data) in which the first column

Table 5.12   Contingency table of the sensory analysis in the survey of the level of liking to drink wine.

| Sample | Evaluation (level of liking) | | | | | | | |
|---|---|---|---|---|---|---|---|---|
| | 1 | 2 | 3 | 4 | 5 | 6 | 7 | |
| Males | 13 | 15 | 18 | 40 | 54 | 57 | 40 | 237 |
| Females | 19 | 20 | 32 | 25 | 28 | 18 | 7 | 149 |
| | 32 | 35 | 50 | 65 | 82 | 75 | 47 | 386 |

contains the labels of the groups (denoted by 1 and 2) and the second contains the observations; the number of permutations (B) and the type of alternative hypothesis (alt) that can be 1 or $-1$ for upper- and lower-tailed alternatives, respectively, and 0 for the two-sided test. The matrix of data in $R$ should be:

```
    Groups Obs
1        1   4
2        1   6
...     ... ...
237      1   7
238      2   3
239      2   5
...     ... ...
386      2   4
```

The observed contingency table is shown in Table 5.12.

By denoting with $Evaluation_M$ and $Evaluation_F$ the ordered categorical response variables, representing the evaluation of males and females, respectively, the system of hypotheses can be formalized as:

$$H_0 : Evaluation_M \overset{d}{=} Evaluation_F$$

against

$$H_1 : Evaluation_M \overset{d}{>} Evaluation_F.$$

The $R$ code for the analysis is as follows:

```
> source("ad.r")
> source("t2p.r")
> source("ad_perm.r")
> wine=read.csv("wine.csv",sep=";",header=TRUE)
> wine[,1]=rep(1:2,c(237,149))
> ad.perm(wine,B=10000,alt=1)
```

The final output is:

```
##########################################################################
# Anderson--Darling Permutation Test:
# Observed Statistic:   3.159619
# p.value:
# [1] 0.0000999
##########################################################################
```

Before the performance of the test, some basic commands are needed. With the source command the codes "ad.r" (for the computation of the Anderson–Darling type test statistic), "t2p.r" (for the computation of the p-value) and "ad_perm.r" (for the application of the permutation test) are loaded. With the read.csv instruction the dataset is imported from the file wine.csv and stored in variable wine. In the dataset the first 237 rows must correspond to the observations of the first sample (males) and the remaining 149 to the observation of the second sample (females). Then the first column of the dataset denotes the sample: hence the label must be 1 for the first 237 rows and 2 for the others. The *p*-value is less than 0.0001 and the null hypothesis should be rejected in favor of the alternative hypothesis that the evaluation of males tends to be better than that of females.

Let us consider another example related to a survey about habits and behaviors of students at the University of Ferrara in Italy. Two samples of students, (1) enrolled for the first year and (2) other students, were asked how often they attend the university facilities (classrooms, libraries, etc.). The goal is to test whether the students enrolled for the first year in the university system tend to attend the facilities more often than other students at the significance level $\alpha = 0.05$. By denoting with *Attendance$_I$* and *Attendance$_{II}$* the ordered categorical response variables, corresponding to the group of students enrolled for the first year and to the group of other students, respectively, the test of hypothesis can be formalized as follows:

$$H_0 : Attendance_I \overset{d}{=} Attendance_{II} \text{ against } H_1 : Attendance_I \overset{d}{>} Attendance_{II}.$$

The ordered categories 'Never', 'Few times a year', 'Few times a month', 'Once or twice a week', 'Three to five times a week', 'Six or seven times a week' are represented in the dataset by the labels '1' to '6'. The dataset must be prepared as follows:

```
   Groups Obs
1      1   5
2      1   5
...  ... ...
123    1   6
124    2   5
125    2   5
...  ... ...
676    2   5
```

Table 5.13   Contingency table of the frequency of attending university facilities by students enrolled for the first year and by other students.

| Group of students | Frequency | | | | | | |
|---|---|---|---|---|---|---|---|
| | 1 | 2 | 3 | 4 | 5 | 6 | |
| 1st year | 2 | 2 | 7 | 4 | 91 | 17 | 123 |
| Others | 30 | 27 | 41 | 48 | 395 | 12 | 553 |
| | 32 | 29 | 48 | 52 | 486 | 29 | 676 |

The contingency table for this problem is shown in Table 5.13.

With a syntax similar to the previous example, considering that data must be imported from the file library.csv, the R code for this analysis is:

```
> source("ad.r")
> source("t2p.r")
> source("ad_perm.r")
> library=read.csv("library.csv",sep=";",header=TRUE)
> library[,1]=rep(1:2,c(123,553))
> ad.perm(library,B=10000,alt=1)
```

The result is the following:

```
######################################################################
# Anderson--Darling Permutation Test:
# Observed Statistic:   5.354041
# p.value:
# [1] 0.000999001
######################################################################
```

The $p$-value is far less than $\alpha$, hence we reject the null hypothesis in favor of the alternative that the students enrolled for the first year at university attend the facilities more often than other types of students.

## 5.4.2   Permutation test on moments

An alternative solution to the problems of goodness-of-fit described in Section 5.4.1 consists of transforming the ordinal data into numerical scores. Each category $A_k$ is replaced by a score $\omega_k$ such that, for each couple $(r, s)$ with $A_r \prec A_s$, the strict inequality $\omega_r < \omega_s$ is true. By considering that two discrete distributions defined on the same support, with a finite number $C$ of distinct real values, are equal if and only if their first $C - 1$ moments are equal, the testing problem for stochastic dominance alternative can be defined as

$$H_0 : \bigcap_{r=1}^{C-1} [\mu_{r,1} = \mu_{r,2}] \text{ against } H_1 : \bigcup_{r=1}^{C-1} [\mu_{r,1} > \mu_{r,2}],$$

where $\mu_{r,j}$ denotes the moment of order $r$ of the transformed variable for the $j$th population, the symbol $\bigcap$ indicates that the null hypothesis is true if all the $C - 1$ sub-hypotheses related to the equality of moments are true and symbol $\bigcup$ denotes that the alternative hypothesis is true if at least one alternative sub-hypothesis (inequality of moments) is true. Note that, in this special case of directional alternative, $H_1$ is equivalent to stochastic dominance only if we assume that under $H_1$ the response variable in the first population is equal to the response of the second plus a non-negative (random or fixed) effect (Arboretti and Bonnini, 2008). This way of representing the testing problem as a set of component-wise hypotheses is used within the multivariate and multi-aspect permutation tests.

For each $r$, a permutation 'partial' test can be performed with test statistic

$$T_r^* = \sum_{k=1}^{C} \omega_k^r \left( f_{1k}^* - f_{2k}^* \right)$$

and, in order to obtain the test statistic for the global problem, the NPC methodology might be applied to combine the $C - 1$ partial tests into a global test. This is achieved by combining the significance level functions of the partial tests with a suitable combining function and then computing the $p$-value of the new univariate combined test statistic.

The permutation test on moments is preferable with respect to the Anderson–Darling type permutation test when the ordered categories can be replaced by a reasonable set of scores. Examples of this are when a Likert scale is used or in general when the response variable represents satisfaction judgments associated with numerical scores from the original questionnaire administered to the respondents. Arboretti and Bonnini (2009) proved that this test has a good power behavior expecially when a symmetric or quite symmetric score transformation is applied.

In the example of sensory analysis of wine (Table 5.12 ), where the goal is to test whether the level of liking of males is higher than that of females, the permutation test on moments can be applied. The ordinal categories can be replaced by their respective ranks as: 'dislike extremely'=1, 'dislike very much'=2, 'dislike slightly'=3, 'neutral'=4, 'like slightly'=5, 'like very much'=6 and 'like extremely'=7.

The function moments.perm(data,B,fun,alt) performs the test. This function requires the following input: the matrix of data (data) in which the first column contains the labels of the group (denoted with 1 and 2) and the second contains the observations of the response; the number of permutations (B), the combination function (fun) with the possibility to choose among "Fisher", "Liptak" and "Tippet" and the type of alternative hypothesis (alt): "greater", "less" or "two.sided". The default values are `fun="Fisher"` and `alt="greater"`. The $R$ code for the analysis and the output are:

```
> source("t2p.r")
> source("combine.r")
> source("moments_perm.r")
> wine=read.csv("wine.csv",sep=";",header=TRUE)
> wine[,1]=rep(1:2,c(237,149))
> moments.perm(wine,B=10000,fun="Fisher",alt="greater")
######################################################################
# Permutation test on moments based on k-1 = 6 partial test
# Global p-value:
# [1] 0.00009975
######################################################################
```

Four $R$ files must be loaded: the usual `"t2p.r"` for computing the significance level function and the $p$-values, `"perm_2samples.r"` for the permutation distribution of the partial tests and `"combine.r"` for the nonparametric combination, in addition to the file `"moments_perm.r"` for the computation of the test statistic. For this analysis, we choose the Fisher combining function and, according to the global $p$-value almost 0, the null hypothesis must be rejected as in the previous analysis based on the Anderson–Darling permutation test.

Let us apply now the permutation test on moments to the problem of the survey on habits and behaviors of the students at the University of Ferrara (Table 5.13). The goal is to test whether the students enrolled for the first year attend the facilities more often than other students. In this case, the categories are transformed as follows: 'Never'=1, 'Few times a year'=2, 'Few times a month'=3, 'Once or twice a week'=4, 'Three to five times a week'=5 and 'Six or seven times a week'=6. By following the same steps as before, the $R$ code and the output are:

```
> source("t2p.r")
> source("combine.r")
> source("moments_perm.r")
> library=read.csv("library.csv",sep=";")
> library[,1]=rep(1:2,c(123,553))
> moments.perm(library,B=10000,fun="Fisher",alt="greater")
######################################################################
# Permutation test on moments based on k-1 = 5 partial test
# Global p-value:
# [1] 0.000999001
######################################################################
```

Thus the resulting $p$-value leads to rejecting the null hypothesis of no difference between the two groups in favor of the alternative that the students enrolled for the first year in the university system attend university facilities more often than other students.

## 5.4.3   The chi-square permutation test

For the testing problem

$$H_0 : \pi_{jk} = \pi_{j\bullet} \times \pi_{\bullet k}, \quad j = 1, \dots, R, \quad k = 1, \dots, C$$

against

$$H_1 : H_0 \text{ not true,}$$

the chi-square test has to be used, provided that each expected frequency is roughly greater than 5 so as to use the asymptotic distribution, conditional on marginal frequencies. However, it should be noted that, due to the conditioning on marginal frequencies (the set of sufficient statistics in $H_0$ for the problem), the true conditional distribution of the chi-square test statistic is exactly the one provided by permutation arguments.

For non-dominance alternatives and nominal or non-ordered categorical variables, a permutation solution for the testing problems related to $R \times C$ contingency tables might be based on the Pearson's chi-square test statistic:

$$T^* = \sum_{j=1}^{R} \sum_{k=1}^{C} \frac{\left( f_{jk}^* - \dfrac{f_{j\bullet} f_{\bullet k}}{n} \right)^2}{\left( \dfrac{f_{j\bullet} f_{\bullet k}}{n} \right)},$$

where $f_{jk}^*$ is the absolute frequency of the $j$th row and $k$th column of the contingency table after a permutation and $f_{j\bullet}, f_{\bullet k}$ and $n$ are the marginal frequencies as usual. The observed value of the test statistic is computed by applying the formula with $f_{jk}$ instead of $f_{jk}^*$. This statistic can be used both for the $R$-sample problem in the presence of a categorical variable taking $C$ possible categories and for the one-sample problem on independence between two categorical variables taking $R$ and $C$ possible categories, respectively.

For example, let us apply the testing procedure to the data of the survey on students performed in Ferrara where some groups of students (defined according to their housing situation) were asked to say how often they used to have lunch in a bar or restaurant in Ferrara. The goal consists in testing whether the frequency of lunch in a bar or restaurant in Ferrara depends on the housing situation of students ($\alpha = 0.01$).

The categorical variable *Housing* represents the housing situation of students and its categories are:

- *Residents*: residents in the city, that is students who reside in the city permanently, not only for the purpose of studying.

- *Offsite students*: students who live in the city for the purpose of studying in specific periods (for attending lectures, doing exams, etc.) but who reside in another place.

- *Commuters*: students who do not reside and do not live in the city but daily or almost daily commute there for the purpose of attending lectures, doing exams, etc.

- *Non-attending students*: students who do not reside and do not live in the city and rarely commute there for the purpose of attending lectures, doing exams, etc.

The categorical variable *Frequency of lunch* represents the frequency of lunch in a bar or restaurant in Ferrara and the categories are:

- Never;
- Few times a year;
- Few times a month;
- Once or twice a week;
- 3–5 times a week.

The corresponding $5 \times 4$ contingency table is shown in Table 5.14.

Formally, we wish to test the null hypothesis of independence against the alternative hypothesis of dependence between frequency of lunch and housing condition. Even if *Frequency of lunch* is an ordered variable, in this problem it is considered like a nominal one. According to the goal of the problem, the ordering of categories is not a useful information because we are not testing for a stochastic ordering (for which we should use an Anderson–Darling test statistic). As a matter of fact the chi-square statistic on contingency tables does not take into consideration the ordered nature of the data and so it is not appropriate for directional alternatives.

Table 5.14   Contingency table of the frequency of lunch in bars or restaurants in Ferrara by different groups of students enrolled in the local university, grouped by housing condition.

| Frequency of lunch | Housing condition | | | | |
|---|---|---|---|---|---|
| | Residents | Offsite | Commuters | Non-attending | |
| Never | 41 | 97 | 71 | 68 | 277 |
| Few times a year | 8 | 20 | 15 | 21 | 64 |
| Few times a month | 18 | 57 | 18 | 28 | 121 |
| Once or twice a week | 44 | 69 | 5 | 12 | 175 |
| 3–5 times a week | 9 | 38 | 56 | 7 | 110 |
| | 120 | 281 | 210 | 136 | 747 |

The function `chisq.test(x,simulate.p.value,B)` performs the chi-square test for contingency tables. The function requires the contingency table represented by a matrix x with $R$ rows and $C$ columns with $R, C \geq 2$. When the second argument is `simulate.p.value=TRUE` it is possible to compute $p$-values by Monte Carlo simulation with $B$ replicates, that is to apply the nonparametric solution where the null distribution of the test statistic is computed by resampling the data $B$ times and holding fixed the marginal frequencies of the table. If `simulate.p.value=FALSE`, the $p$-value is computed according to the asymptotic chi-square distribution of the test statistic with $(R - 1) \cdot (C - 1)$ degrees of freedom. The $R$ code for the analysis is:

```
> data=matrix(c(41,97,71,68,8,20,15,21,18,57,18,28,44,69,50,
+ 12,9,38,56,7),ncol=4,byrow=TRUE)
> chisq.test(data,simulate.p.value=TRUE,B=10000)
######################################################################
# Pearson's Chi-squared test with simulated p-value
# (based on 10000 replicates)
#
# data:   data
# X-squared = 83.7185, df = NA, p-value = 0.0000999
######################################################################
```

The function returns the observed value of the chi-square test statistic, the degrees of freedom and the $p$-value. In this case, the degrees of freedom are not available because we use the permutation test and not the parametric one based on the asymptotic chi-square distribution. Note that the resulting $p$-value is less than $\alpha$, thus the null hypothesis of independence must be rejected in favor of the alternative hypothesis that the frequency of having lunch in a bar or restaurant in Ferrara depends on the housing situation of students.

Table 5.15   Contingency table of the type of activity by age group in domestic accidents.

| Activity | Age group (yr) | | | |
| --- | --- | --- | --- | --- |
| | 0–24 | 25–64 | >64 | |
| Personal care | 2 | 5 | 2 | 9 |
| Housework | 1 | 26 | 15 | 42 |
| Repairing objects | 0 | 6 | 0 | 6 |
| Playing games | 8 | 1 | 2 | 11 |
| None | 3 | 11 | 3 | 17 |
| Other | 1 | 1 | 3 | 5 |
| | 15 | 50 | 25 | 90 |

The second application is related to a survey about domestic accidents in which a sample of people of different age groups who had accidents at home was asked to indicate the type of activity they were doing at the moment of the accident. Three age groups (in years) were considered: 0–24, 25–64 and > 64 (over 64). The types of activity were: personal care, housework, repairing objects, playing games, none, and other. The goal of the study is to test whether the type of activity depends on age at the significance level $\alpha = 0.05$. The corresponding $6 \times 3$ contingency table is shown in Table 5.15.

Similarly to the previous example, the R code allows to assign the observed frequencies to variable data and to perform the analysis with the chi-square permutation test is as follows:

```
> data=matrix(c(2,5,2,1,26,15,0,6,0,8,1,2,3,11,3,1,1,3),
+ ncol=3,byrow=TRUE)
> chisq.test(data,simulate.p.value=TRUE,B=10000)
######################################################################
# Pearson's Chi-squared test with simulated p-value
# (based on 10000 replicates)
#
# data:  data
# X-squared = 40.7826, df = NA, p-value = 0.0000999
######################################################################
```

The resulting $p$-value is far less than $\alpha = 0.05$, hence the null hypothesis is rejected in favor of the hypothesis of dependence between type of activity and age group.

# References

Anderson, T.W. and Burstein, H. (1967) Approximating the upper binomial confidence limit. Journal of the American Statistical Association, 62, 857–861.

Anderson, T.W. and Burstein, H. (1968) Approximating the lower binomial confidence limit. Journal of the American Statistical Association, 63, 1413–1415.

Arboretti Giancristofaro, R. and Bonnini, S. (2008) Moment-based multivariate permutation tests for ordinal categorical data. Journal of Nonparametric Statistics, 20, 383–393.

Arboretti Giancristofaro, R. and Bonnini, S. (2009) Nonparametric directional tests in presence of confounding factors and categorical data. Statistica & Applicazioni, VII, 87–103.

Bennett, B.M. and Underwood, R.E. (1970) On McNemar's test for the 2×2 table and its power function. Biometrics, 26, 339–343.

Clopper, C.J. and Pearson, E.S. (1934) The use of confidence or fiducial limits illustrated in the case of the binoial. Biometrika, 26, 404–413.

Cohen, A. and Sackrowitz, H.B (1998) Directional tests for one-sided alternatives in multivariate models. Annals of Statistics, 26, 2321–2338.

Conover, W.J. (1999) Practical Nonparametric Statistics. John Wiley & Sons, Ltd.

El Barmi, H. and Dykstra, R. (1995) Testing for and against a set of linear inequality constraints in a multinomial setting. Canadian Journal of Statistics, 23, 131–143.

El Barmi, H. and Mukerjee, H. (2005) Inferences under stochastic ordering constraint: the k-sample case. Journal of the American Statistical Association, 100, 252–261.

Goodman, L.A. (1965) On simultaneous confidence intervals for multinomial proportions. Technometrics, 7, 247–254.

Kvam, P.H. and Vidakovic, B. (2007) Nonparametric Statistics with Applications to Science and Engineering. John Wiley & Sons, Ltd.

Mantel, N. and Fleiss, J.L. (1975) The equivalence of the generalized McNemar tests for marginal homogeneity in $2^3$ and $3^2$ tables. Biometrics, 31, 727–729.

McKinlay, S.M. (1975) A note on the chi-square test for pair-matched samples. Biometrics, 31, 731–735.

McNemar, Q. (1947) A Note on the sampling error of the difference between correlated proportions and percentages. Psychometrika, 12, 153–157.

Paek, S.Y., Steimberg, S.M. and Katz, S.I. (2011) Remission in dermatitis herpetiformis: a cohort study. Archives of Dermatology, 147, 301–305.

Pesarin, F. (2001) Multivariate Permutation Tests with Applications in Biostatistics. John Wiley & Sons Ltd.

Pesarin, F. and Salmaso, L. (2010) Permutation Tests for Complex Data: Theory, Applications and Software. John Wiley & Sons, Ltd.

Quesenberry, C.P. and Hurst, D.C. (1964) Large sample simultaneous confidence intervals for multinomial proportions. Technometrics, 6, 191–195.

Sampson, A.R. and Whitaker, L.R. (1989) Estimation of multivariate distributions under stochastic ordering. Journal of the American Statistical Association, 84, 541–548.

Ury, H.K. (1975) Efficiency of case–control studies with multiple controls per case: continuous or dichotomous data. Biometrics, 3, 643–650.

Wang, Y. (1996) A likelihood ratio test against stochastic ordering in several populations. Journal of the American Statistical Association, 91, 1676–1683.

# 6

# Testing for correlation and concordance

## 6.1 Introduction

In this chapter, we study correlation and concordance. This section introduces the problems to be addressed and the method presented in the successive sections.

In Section 6.2 we are interested in the statistical relationship between the two components $X$ and $Y$ of the bivariate variable underlying the population of interest. The objective is to understand whether or not these variables are independent and in case they are not independent to assess the degree of dependency among them. The Pearson product moment correlation coefficient, that is the most familiar measure of correlation, is introduced. Its distribution under the null hypothesis that $X$ and $Y$ are independent depends on the distribution of the bivariate variable $(X, Y)$. For this reason it is not suitable as the test statistic within a nonparametric framework for testing the hypothesis that $X$ and $Y$ are independent.

In Section 6.3 we consider nonparametric tests for independence. More precisely, in Section 6.3.1 we consider the nonparametric test based on the Spearman correlation coefficient and in Section 6.3.2 we consider the nonparametric test based on the Kendall correlation coefficient. Both tests are based on ranks and the null distributions of the corresponding test statistics do not depend on the distribution of $(X, Y)$: the tests are distribution free. The only assumption is that $X$ and $Y$ are continuous. Applications to real life problems in finance, management and experimental education are discussed. More precisely, we study the correlation between functional dispersion and hierarchical control for a sample of manufacturing firms, the correlation between the net profit to staff hours ratio and the net profit to sale space ratio for a group of hypermarkets, and the correlation between the GMAT (Graduate Management

*Nonparametric Hypothesis Testing: Rank and Permutation Methods with Applications in R*,
First Edition. Stefano Bonnini, Livio Corain, Marco Marozzi and Luigi Salmaso.
© 2014 John Wiley & Sons, Ltd. Published 2014 by John Wiley & Sons, Ltd.
Companion website: http://www.wiley.com/go/hypothesis_testing

Admission Test) score taken before entering graduate school and the grade point average while in the MBA (Master in Business Administration) program for a sample of graduates.

In Section 6.4 the problem of whether the rankings of some objects given by a set of criteria (or judges) show any agreement or are independent is addressed. Rankings of a group of objects are very often considered in practice when for example job applicants, new products, services and investments are ranked by head hunters, focus groups, investors or even an automated algorithm. The most familiar measure for concordance is the Kendall $W$ coefficient. Classical tests for concordance are the Friedman test and the $F$ test due to Kendall and Babington–Smith. Section 6.4.1 presents the Kendall and Babington–Smith $F$ test and Section 6.4.2 presents a permutation test for concordance. An application to finance is discussed. More precisely we assess an important financial problem: to understand whether a set of firm financial ratios is concordant or not.

Sections 6.2 and 6.3

Data: $n$ bivariate pairs $\{(X_i, Y_i), i = 1, \dots, n\}$.

Assumption

A: $\{(X_i, Y_i), i = 1, \dots, n\}$ is an i.i.d. random sample from a continuous bivariate population.

Section 6.4

Data: ranks of $n$ objects given by (according to) $p$ judges (criteria) $R_{ij} = \mathbb{R}(X_{ij})$; $i = 1, \dots, n; j = 1, \dots, p$.

Assumption

B: $X_j, j = 1, \dots, p$ is a random variable to be ranked.

# 6.2    Measuring correlation

In this section, we consider a bivariate population and we would like to assess the statistical relationship between the two components $X$ and $Y$ of the bivariate variable underlying the population of interest. The objective is to understand whether or not these variables are independent and in case they are not independent to assess the degree of dependency among them. The requirements of a measure of correlation to be acceptable are reported and the most familiar measure of correlation, that is the Pearson product moment correlation coefficient, is introduced.

Reimann (1974) studied the organization structure of a set of 19 northeast Ohio manufacturing firms. Data were collected by interviewing top executives of each firm as well as by consulting organization charts, standard operating procedures and policy manuals. Among other data, the scores of the firms on two effectiveness criteria have been collected: functional dispersion $X$ and hierarchical control $Y$. Functional dispersion represents the extent to which individuals are evenly distributed among the various speciality functions in the firm. When all the functions of the firm have roughly the same number of individuals the score is high. When the bulk of individuals is concentrated in one or two functions and in each of the other functions there are one or two individuals, the score is low. Hierarchical control is the degree of cumulative authority and responsibility resting in the various levels of firm hierarchy. The score

Table 6.1   Scores on functional dispersion $X$ and hierarchical control $Y$ for a sample of manufacturing firms.

| Company | 1 | 2 | 3 | 4 | 5 | 6 | 7 | 8 | 9 | 10 |
|---|---|---|---|---|---|---|---|---|---|---|
| $X$ | 0.812 | 0.716 | 0.717 | 0.885 | 0.891 | 0.930 | 0.505 | 0.840 | 0.863 | 0.819 |
| $Y$ | 23.8 | 7.6 | 5.6 | 23.2 | 8.8 | 13.0 | 25.1 | 14.2 | 8.5 | 25.2 |

| Company | 11 | 12 | 13 | 14 | 15 | 16 | 17 | 18 | 19 |
|---|---|---|---|---|---|---|---|---|---|
| $X$ | 0.468 | 0.907 | 0.538 | 0.830 | 0.800 | 0.760 | 0.376 | 0.690 | 0.690 |
| $Y$ | 6.4 | 5.7 | 1.9 | 12.2 | 3.8 | 13.5 | 5.0 | 3.9 | 8.8 |

is minimum when the chief executive has all the other employees reporting directly to him/her, that is the firm has the flattest possible structure. The score is maximum when every individual except the bottom one has one immediate subordinate, that is the firm has the tallest possible structure. Table 6.1 reports the $X$ and $Y$ score. We would like to measure the correlation between the two variables.

To be acceptable a measure of correlation between $X$ and $Y$ should satisfy the following properties:

1. The measure of correlation should lay between $-1$ and 1.

2. If the larger values of $X$ and $Y$ tend to be paired together and so are the smaller values then the measure of correlation should be positive and the stronger the tendency the closer to 1. In this case we say that $X$ and $Y$ are positively correlated (some authors would say that $X$ and $Y$ are concordant).

3. If the larger values of $X$ tend to be paired with the smaller values of $Y$, and the smaller values of $X$ tend to be paired with the larger ones of $Y$ then the measure of correlation should be negative and the stronger the tendency the closer to $-1$. In this case we say that $X$ and $Y$ are negatively correlated (some authors would say that $X$ and $Y$ are discordant).

4. If $X$ and $Y$ are neither positively correlated nor negatively correlated the values of $X$ are randomly paired with the values of $Y$ and then the measure of correlation should be close to 0. In this case we say that $X$ and $Y$ are uncorrelated or linearly independent.

The most familiar measure of correlation is the Pearson product moment correlation coefficient

$$r = \frac{\sum_{i=1}^{n}(X_i - \overline{X})(Y_i - \overline{Y})}{\left[\sum_{i=1}^{n}(X_i - \overline{X})^2 \sum_{i=1}^{n}(Y_i - \overline{Y})^2\right]^{1/2}}$$

$$= \frac{\sum_{i=1}^{n} X_i Y_i - n\overline{X}\overline{Y}}{\left(\sum_{i=1}^{n} X_i^2 - n\overline{X}^2\right)^{1/2}\left(\sum_{i=1}^{n} Y_i^2 - n\overline{Y}^2\right)^{1/2}},$$

which is the ratio between the sample covariance and the product of the sample standard deviations (by dividing both numerator and denominator by $n$ the ratio does not change). This is a measure of the strength of the linear association between $X$ and $Y$. If we plot $Y$ versus $X$, the closer the points $(X_i, Y_i)$, $i = 1, \ldots, n$ to a straight line the closer $r$ to 1 if the line is sloping upward and to $-1$ if the line is sloping downward. Note that the square of $r$ measures the quality of a least squares fitting to the original data. It is important to emphasize that the assumptions reported above are satisfied. As a descriptive statistic, $r$ may be used with all numeric data and without any assumptions on the shape of the underlying distributions of the $(X, Y)$ random variable.

It is very easy to compute the Pearson correlation coefficient using $R$, just load the data of Table 6.1

```
> #functional dispersion
> x=c(.812,.716,.717,.885,.891,.930,.505,.840,.863,.819,.468,
.907,.538,.830,.800,.760,.376,.690,.690)
>
> #hierarchical control
> y=c(23.8,7.6,5.6,23.2,8.8,13.0,25.1,14.2,8.5,25.2,6.4,5.7,1.9,
12.2,3.8,13.5,5.0,3.9,8.8)
```

and run

```
> cor(x,y)
[1] 0.2411785
```

We see that the strength of the linear positive association between functional dispersion and hierarchical control for a sample of manufacturing firms is 24, 1% of the maximum possible strength. To understand whether the evidence of correlation is strong enough to not be due to chance we should perform a test of the null hypothesis that $X$ and $Y$ are uncorrelated against the alternative hypothesis that $X$ and $Y$ are positively correlated. Unfortunately, $r$ cannot be used as a pivotal statistic for a nonparametric test because under the null hypothesis its distribution depends on the bivariate distribution of $(X, Y)$. Two solutions to this problem are presented in Section 6.3. The first one is based on the Spearman correlation coefficient and the second one is based on the Kendall correlation coefficient.

## 6.3   Tests for independence

In this section we consider two measures of correlation, due to Spearman and Kendall. Contrary to the Pearson product moment correlation coefficient they have distribution functions that are independent of the distribution function of $(X, Y)$ if $X$ and $Y$ are independent and continuous. Therefore these measures of correlation can be used as test statistics for nonparametric testing the null hypothesis that $X$ and $Y$ are independent. With some caution, they may even be used for ordered categorical data.

## 6.3.1  The Spearman test

Let $\mathbb{R}(X_i)$ be the rank of $X_i$ compared with the other $X$ values for $i = 1, \dots, n$ and similarly let $\mathbb{R}(Y_i)$ be the rank of $Y_i$. It is important to note that $X$ and $Y$ may even be non numerical variables, provided that they can be ranked, that is that they are at least ordered categorical. The ranking of an ordered categorical variable may be based on the quality of the observations or the degree of preference attached to the observations. Note that in certain situations, even though the variable is numerical, it may be of greater interest to consider the rank rather than the numerical value. An example is the United Nations Human Development Index. Many scholars prefer to study and compare the ranking of nations, among a certain group of them or across time, on the Human Development Index rather than on the value of the index itself.

The Spearman correlation coefficient is nothing else than the Pearson correlation coefficient applied to the ranks $(\mathbb{R}(X_i), \mathbb{R}(Y_i))$ rather than to the original observations $(X_i, Y_i)$, $i = 1, \dots, n$.

$$\rho = \frac{\sum_{i=1}^{n} (\mathbb{R}(X_i) - \overline{\mathbb{R}}(X))(\mathbb{R}(Y_i) - \overline{\mathbb{R}}(Y))}{\left[ \sum_{i=1}^{n} (\mathbb{R}(X_i) - \overline{\mathbb{R}}(X))^2 \sum_{i=1}^{n} (\mathbb{R}(Y_i) - \overline{\mathbb{R}}(Y))^2 \right]^{1/2}}$$

$$= \frac{\sum_{i=1}^{n} \mathbb{R}(X_i)\mathbb{R}(Y_i) - n\overline{\mathbb{R}}(X)\overline{\mathbb{R}}(Y)}{\left( \sum_{i=1}^{n} (\mathbb{R}(X_i))^2 - n(\overline{\mathbb{R}}(X))^2 \right)^{1/2} \left( \sum_{i=1}^{n} (\mathbb{R}(Y_i))^2 - n(\overline{\mathbb{R}}(Y))^2 \right)^{1/2}},$$

where $\overline{\mathbb{R}}(X) = \overline{\mathbb{R}}(Y) = (n + 1)/2$ denotes the average rank of $X$ and $Y$. In the case of tied observations assign to each tie the mid rank, that is the average of the ranks that would have been assigned if there had been no ties. In the absence of ties, an equivalent and computationally easier formula for the Spearman coefficient is

$$\rho = 1 - \frac{6 \sum_{i=1}^{n} \left[ \mathbb{R}(X_i) - \mathbb{R}(Y_i) \right]^2}{n(n^2 - 1)}.$$

Note that this formula was generally used before the widespread of computers. Today it is of interest particularly for educational purposes.

Using $R$ the computation of the Spearman correlation coefficient is as easy as for the Pearson correlation coefficient. To compute the Spearman correlation coefficient between functional dispersion and hierarchical control run

```
> #functional dispersion
> x=c(.812,.716,.717,.885,.891,.930,.505,.840,.863,.819,.468,.907,
.538,.830,.800,.760,.376,.690,.690)
> #hierarchical control
> y=c(23.8,7.6,5.6,23.2,8.8,13.0,25.1,14.2,8.5,25.2,6.4,5.7,1.9,
12.2,3.8,13.5,5.0,3.9,8.8)
> cor(x,y,method="spearman")
[1] 0.3472344
```

Let $F_{X,Y}$ denote the joint distribution function of the bivariate random variable $(X, Y)$, let $F_X$ and $F_Y$, denote the marginal distribution function of $X$ and $Y$, respectively. The null hypothesis of interest is that $X$ and $Y$ are mutually independent

$$H_0 : \{F_{X,Y}(x, y) = F_X(x) F_Y(y), \forall (x, y) \in \mathcal{R}^2\}.$$

The alternative hypotheses of interest correspond to the concepts of positive and negative correlation between $X$ and $Y$.

(i) One-sided test (positive correlation). To test the null hypothesis of independence between $X$ and $Y$ against the one-sided alternative hypothesis

$$H_1 : \{X \text{ and } Y \text{ are positively correlated}\},$$

that is the larger values of $X$ tend to be paired with the larger values of $Y$ and the smaller values of $X$ tend to be paired with the smaller values of $Y$.

(ii) One-sided test (negative correlation). To test the null hypothesis of independence between $X$ and $Y$ against the one-sided alternative hypothesis

$$H_1 : \{X \text{ and } Y \text{ are negatively correlated}\},$$

that is the larger values of $X$ tend to be paired with the smaller values of $Y$ and the smaller values of $X$ tend to be paired with the larger values of $Y$.

(iii) Two-sided test. To test the null hypothesis of independence between $X$ and $Y$ against the two-sided alternative hypothesis

$$H_1 : \{(X \text{ and } Y \text{ are positively correlated}) \text{ or } (X \text{ and } Y \text{ are negatively correlated})\}.$$

The test statistic is always

$$S = \sum_{i=1}^{n} [\mathbb{R}(X_i) - \mathbb{R}(Y_i)]^2.$$

It is worth noting that ranks are one-to-one related with empirical distribution functions: $\mathbb{R}(X_i) = n\widehat{F}_X(X_i)$ and $\mathbb{R}(Y_i) = n\widehat{F}_Y(Y_i)$. Of course the $p$-value of the corresponding test differs according to the alternative hypothesis of interest. In case (i) the $p$-value is the probability that $S$ is less than or equal to its observed value. It should be emphasized that in this case the test rejects for small values of $S$ because when the alternative hypothesis of positive correlation is true $\mathbb{R}(X_i)$ and $\mathbb{R}(Y_i)$, $i = 1, \ldots, n$ tend to agree and the differences $\mathbb{R}(X_i) - \mathbb{R}(Y_i)$, $i = 1, \ldots, n$ tend to be small. Therefore it is expected that $S$ is small and the observed Spearman correlation coefficient large and positive. Note that $\rho$ and $S$ are one-to-one decreasingly related with

$$S = \frac{(1 - \rho)n(n^2 - 1)}{6}.$$

In case (ii) the $p$-value is 1 minus the $p$-value of case (i) since it is the opposite case: when the alternative hypothesis of negative correlation is true, $\mathbb{R}(X_i)$ and $\mathbb{R}(Y_i)$, $i = 1, \ldots, n$ tend to disagree and the differences $\mathbb{R}(X_i) - \mathbb{R}(Y_i)$, $i = 1, \ldots, n$ tend to be large. Therefore it is expected that $S$ is large and the observed Spearman correlation coefficient large and negative. In case (iii) the $p$-value is 2 times the $p$-value of case (i) since it is the two-sided case. Note that the distribution of the $S$ statistic under the null hypothesis is symmetric. Several methods to compute the $p$-value of the Spearman test have been proposed in the literature. The basic package of $R$ uses the algorithm of Best and Roberts (1975) which computes an exact $p$-value for $n < 9$ in the absence of ties. When $10 < n < 1290$ an Edgeworth series approximation is used. In the other cases an asymptotic approximation is used. Note that in the literature different test statistics for the Spearman test have been proposed. All of them are linear functions of $\sum_{i=1}^{n} \mathbb{R}(X_i)\mathbb{R}(Y_i)$ and then they are equivalent for the purpose of hypothesis testing.

The $R$ code for testing that the Spearman correlation coefficient between functional dispersion and hierarchical control is significantly different than 0 (two-sided test) is as follows:

```
> cor.test(x,y,alternative="two.sided",method="spearman")
####################################################################
# Spearman's rank correlation rho
#data: x and y
#S = 744.1528, p-value = 0.1452
#alternative hypothesis: true rho is not equal to 0
#sample estimates:
# rho
#0.3472344
####################################################################
```

Note that $R$ warns about the presence of ties and then computes an approximate $p$-value. Since the $p$-value is 0.145 there is no strong evidence that functional dispersion and hierarchical control of a group of manufacturing firms are positively or negatively correlated. To obtain the $S$ statistic write

```
> cor.test(x,y,alternative="two.sided",method="spearman")$statistic
  S
744.1528
```

and to obtain the corresponding $p$-value write

```
> cor.test(x,y,alternative="two.sided",method="spearman")$p.value
[1] 0.1452317
```

To test the null hypothesis of independence against the one-sided alternative that $X$ and $Y$ are positively correlated just replace alternative="two.sided" with alternative="greater". To test the null hypothesis against the one-sided alternative that $X$ and $Y$ are negatively correlated replace `alternative="two.sided"` with `alternative="less"`. It is worth noting that there are some additional packages that perform the Spearman test within the $R$ environment with different and often more accurate approximations for the $p$-value. For example, the additional $R$ package called `pspearman` allows use of the precompiled null distribution of the $S$ statistic for $n \leq 22$ in the absence of ties. We leave the reader to go into this and other additional packages about the Spearman test thoroughly, since for the purpose of this subsection the basic function it suffices.

It is important to emphasize that the Spearman test is a distribution-free test because the null distribution of $S$ can be obtained without any assumptions on the distributions of the populations behind $X$ and $Y$ except requiring that they be continuous. In the absence of ties, without loss of generality we may assume that $\mathbb{R}(X_{(i)}) = i$, $i = 1, \ldots, n$. Under the null hypothesis, all possible $n!$ permutations of the first $n$ natural numbers are equally likely to be the ranks $\mathbb{R}(Y_i)$, $i = 1, \ldots, n$ and therefore each possible $(\mathbb{R}(X_i), \mathbb{R}(Y_i), i = 1, \ldots, n)$ configuration has probability $1/n!$ under the null hypothesis. Note that, under the null hypothesis, the random variable $\sum_{i=1}^{n} [i - \mathbb{R}(Y_i)]^2$ has the same distribution as $S$. In the presence of ties the null distribution of $S$ is conditional on the data at hand and in particular on the observed sets of tied ranks.

## 6.3.2   The Kendall test

In this subsection we consider another very familiar coefficient of correlation due to Kendall. Like the Spearman correlation coefficient its distribution does not depend on the distribution of $X$ and $Y$ if $X$ and $Y$ are independent and continuous. Therefore it is suitable for nonparametric testing. Prior to the widespread of fast and cheap computers, the Kendall coefficient of correlation was often preferred to the Spearman one because under the null hypothesis of independence, the approximation for the test statistic based on the Kendall coefficient is better (in the sense that it approaches the limit faster) than for the test statistic based on the Spearman coefficient. Nowadays, this computational aspect is less important. Another advantage of the Kendall coefficient, as will be discussed next, which is still important, is its direct and simple interpretation in terms of probabilities of observing concordant and discordant pairs of observations.

We first assume that there are no ties. Two observations $(X_i, Y_i)$ and $(X_j, Y_j)$ are called concordant if $X_i > Y_i$ and $X_j > Y_j$ (or $X_i < Y_i$ and $X_j < Y_j$). They are called discordant if $X_i > Y_i$ and $X_j < Y_j$ (or $X_i < Y_i$ and $X_j > Y_j$). Let $N_c$ denote the number of concordant pairs of observations and let $N_d$ denote the number of discordant pairs.

$$N_c + N_d = \binom{n}{2} = \frac{n(n-1)}{2},$$

which is the total number of ways the $n$ observations may be paired. Note that $N_c$ and $N_d$ can be computed also if we had at our disposal the ranks $(\mathbb{R}(X_i), \mathbb{R}(Y_i)), i = 1, \ldots, n$ rather than the original observations $(X_i, Y_i), i = 1, \ldots, n$. Moreover, the data may also consist of ordered categorical observations because it is only necessary that the concepts of concordant and discordant pairs make sense. The Kendall coefficient of correlation is defined as

$$\tau = \frac{N_c - N_d}{n\,(n-1)\,/2}.$$

When there is tendency neither of concordance nor of discordance $N_c$ and $N_d$ would not be much different and then $\tau$ would be near 0. When observations tend to be concordant $N_c > N_d$ and then $\tau > 0$ with $\tau$ approaching 1 closer and closer as this tendency becomes stronger and stronger. When all pairs of observations are concordant $N_c = n\,(n-1)\,/2, N_d = 0$ and then $\tau = 1$. Conversely, when observations tend to be discordant $N_c < N_d$ and then $\tau < 0$ with $\tau$ approaching $-1$ closer and closer as this tendency becomes stronger and stronger. When all pairs of observations are discordant $N_d = n\,(n-1)\,/2$, $N_c = 0$ and then $\tau = -1$. Note that the Kendall coefficient meets the requirements of a measure of correlation reported in Section 6.2.

To illustrate the computation of the Kendall coefficient of correlation in $R$ we consider the following real life problem (Korhonen and Siljamäki, 1998). In 1990 the largest group of hypermarkets in Finland was a Prisma chain and consisted of 19 large hypermarkets. The hypermarkets were ranked according to some performance indicators such as the ratio $X$ between net profit and staff hours and the ratio $Y$ between net profit and sale space in square meters (Table 6.2).

To compute the Kendall coefficient of correlation run the following code.

```
> #net profit/staff hours, ranks
> x=c(1,2,3,6,4,10,8,15,9,5,7,17,13,12,11,14,16,18,19)
> #net profit/sale space, ranks
> y=c(1,3,2,6,5,7,11,9,13,14,8,12,4,17,16,15,18,10,19)
> cor(x,y,method="kendall")
[1] 0.5321637
```

The Kendall coefficient is equal to 0.532.

Table 6.2   Ranks of a group of hypermarkets according to the ratios $X$ between net profit and staff hours and the ratio $Y$ between net profit and sale space in square meters.

| Hypermarket | 1 | 2 | 3 | 4 | 5 | 6 | 7 | 8 | 9 | 10 |
|---|---|---|---|---|---|---|---|---|---|---|
| $X$ | 1 | 2 | 3 | 6 | 4 | 10 | 8 | 15 | 9 | 5 |
| $Y$ | 1 | 3 | 2 | 6 | 5 | 7 | 11 | 9 | 13 | 14 |

| Hypermarket | 11 | 12 | 13 | 14 | 15 | 16 | 17 | 18 | 19 |
|---|---|---|---|---|---|---|---|---|---|
| $X$ | 7 | 17 | 13 | 12 | 11 | 14 | 16 | 18 | 19 |
| $Y$ | 8 | 12 | 4 | 17 | 16 | 15 | 18 | 10 | 19 |

In the presence of ties, the Kendall coefficient is defined as

$$\tau_b = \frac{N_c - N_d}{\sqrt{(N_c + N_d + t_X)(N_c + N_d + t_Y)}},$$

where $t_X$ is the total number of pairs of observations that are tied in $X$ with

$$t_X = \sum_{l=1}^{L} \binom{t_l}{2},$$

where $L$ is the number of groups of ties in $X$ and $t_l$ is the number of tied $X$ in the $l$th group of ties. Similarly

$$t_Y = \sum_{m=1}^{M} \binom{t_m}{2},$$

where $M$ is the number of groups of ties in $Y$ and $t_m$ is the number of tied $Y$ in the $m$th group of ties. Note that as before each concordant pair adds 1 to $N_c$ whereas each discordant pair adds 1 to $N_d$. Pairs such that $X_i = X_j$ and/or $Y_i = Y_j$ add 0 to both $N_c$ and $N_d$. Note that in the absence of ties $\tau_b$ reduces to $\tau$ since $t_X = t_Y = 0$. The total number of possible pairs in this case is computed as the geometric mean between $(N_c + N_d + t_X)$ and $(N_c + N_d + t_Y)$.

To illustrate the computation of the Kendall coefficient of correlation in the presence of ties, we analyze a dataset taken from Conover (1999). Twelve MBA (Master in Business Administration) graduates are considered. We would like to measure the strength of agreement between their GMAT (Graduate Management Admission Test) score $X$ taken before entering graduate school and their grade point average $Y$ while they were in the MBA program. See Table 6.3.

Table 6.3   GMAT score $X$ and grade point average $Y$ of a group of MBA graduates.

| Graduate | 1 | 2 | 3 | 4 | 5 | 6 |
|---|---|---|---|---|---|---|
| $X$ | 710 | 610 | 640 | 580 | 545 | 560 |
| $Y$ | 4 | 4 | 3.9 | 3.8 | 3.7 | 3.6 |

| Graduate | 7 | 8 | 9 | 10 | 11 | 12 |
|---|---|---|---|---|---|---|
| $X$ | 610 | 530 | 560 | 540 | 570 | 560 |
| $Y$ | 3.5 | 3.5 | 3.5 | 3.3 | 3.2 | 3.2 |

To compute the Kendall coefficient of correlation run the following code.

```
> #GMAT score
> x=c(710,610,640,580,545,560,610,530,560,540,570,560)
>
> #grade point average
> y=c(4.0,4.0,3.9,3.8,3.7,3.6,3.5,3.5,3.5,3.3,3.2,3.2)
>
> cor(x,y,method="kendall")
[1] 0.4390389
```

Note that Conover (1999) uses a different version of the Kendall coefficient in the presence of ties than $\tau_b$ which is used in $R$. To compute explicitly $\tau_b$ in this case, consider that the number of concordant pairs is $N_c = 42$, the number of discordant pairs is $N_d = 15$, the number of tied pairs in $X$ is $t_X = 3 + 1 = 4$ and the number of tied pairs in $Y$ is $t_Y = 3 + 1 + 1 = 5$. Therefore

$$\tau_b = \frac{42 - 15}{\sqrt{(42 + 15 + 4)(42 + 15 + 5)}} = 0.43904.$$

To compute it using $R$ write

```
> (42-15)/sqrt((42+15+4)*(42+15+5))
[1] 0.4390389
```

We conclude that the GMAT score and the grade point average of a group of MBA graduates are positively correlated. To understand whether the practical evidence of $\tau_b = 0.43904$ is strong enough to conclude that the correlation between the GMAT score and the grade point average is significantly greater than 0 we need a test based on the Kendall coefficient of correlation. $\tau$ and $\tau_b$ are point estimates of the Kendall population correlation coefficient

$$\tau_{pop} = 2 \Pr \{\text{a pair of observations is concordant}\} - 1$$
$$= 2 \Pr \{(X_j - X_i)(Y_j - Y_i) > 0\} - 1.$$

When $X$ and $Y$ are perfectly concordant $\tau_{pop} = 2 \cdot 1 - 1 = 1$ whereas when they are perfectly discordant $\tau_{pop} = 2 \cdot 0 - 1 = -1$. When $X$ and $Y$ are independent $\tau_{pop} = 0$. In general it is

$$\{(X_j - X_i)(Y_j - Y_i) > 0\} \Leftrightarrow \{(X_j - X_i > 0) \cap (Y_j - Y_i) > 0)\}$$
$$\cup \{(X_j - X_i < 0) \cap (Y_j - Y_i) < 0)\}.$$

Since it is a union of mutually exclusive events then

$$Pr\{(X_j - X_i)(Y_j - Y_i) > 0\} = Pr\{(X_j - X_i > 0) \cap (Y_j - Y_i) > 0)\}$$
$$+ Pr\{(X_j - X_i < 0) \cap (Y_j - Y_i) < 0)\}.$$

If $X$ and $Y$ are independent then

$$Pr\{(X_j - X_i > 0) \cap (Y_j - Y_i) > 0)\} = Pr\{X_j > X_i\} Pr\{Y_j > Y_i\}$$
$$= \frac{1}{2} \cdot \frac{1}{2} = \frac{1}{4}$$

because $X_i$ and $X_j$ are independent and identically distributed variables, as are $Y_i$ and $Y_j$ thought not necessarily with the same distribution as $X_i$ and $X_j$. Similarly it is

$$Pr\{(X_j - X_i < 0) \cap (Y_j - Y_i) < 0)\} = Pr\{X_j < X_i\} Pr\{Y_j < Y_i\} = \frac{1}{4}.$$

Therefore if $X$ and $Y$ are independent then $\tau_{pop} = 2\left(\frac{1}{4} + \frac{1}{4}\right) - 1 = 0$. Note that the reverse is not always true as $\tau_{pop} = 0$ does not necessarily imply that $X$ and $Y$ are independent.

The null hypothesis of interest is the same as for the Spearman test.

$$H_0 : \{F_{X,Y}(x, y) = F_X(x) F_Y(y), \forall (x, y) \in \mathcal{R}^2\}.$$

Note that the null hypothesis implies that $\tau_{pop} = 0$ and that it is symmetric about its 0 mean. The alternative hypotheses correspond to the concepts of concordance and discordance among pairs of observations.

(i) One-sided upper-tail test. To test the null hypothesis of independence between $X$ and $Y$ against the one-sided upper-tail alternative hypothesis

$$H_1 : \{\text{pairs of observations tend to be concordant}\} = \{\tau_{pop} > 0\}.$$

Large positive values of $\tau_b$ speak against the null hypothesis. The $p$-value is given by the probability that $\tau_b$ test statistic is greater than or equal to the observed $\tau_b$ value.

(ii) One-sided lower-tail test. To test the null hypothesis of independence between $X$ and $Y$ against the one-sided lower-tail alternative hypothesis

$$H_1 : \{\text{pairs of observations tend to be discordant}\} = \{\tau_{pop} < 0\}.$$

Large negative values of $\tau_b$ speak against the null hypothesis. The $p$-value is given by the probability that the $\tau_b$ test statistic is less than or equal to the observed $\tau_b$ value.

(iii) Two-sided test. To test the null hypothesis of independence between $X$ and $Y$ against the two-sided alternative hypothesis

$H_1$ : {pairs of observations either tend to be concordant, or tend to be discordant}
  = $\{\tau_{pop} \neq 0\}$.

Large absolute values of $\tau_b$ speak against the null hypothesis. The $p$-value is given by the probability that $\tau_b$ test statistic in absolute value is greater than or equal to the absolute value of the observed $\tau_b$ value.

Under Assumption A, the test is consistent against the above class of alternatives. It is important to note that in the absence of ties, the Kendall coefficient is a linear function of the simpler $N_c$ statistic which is used in $R$ as the test statistic for hypothesis testing on the Kendall coefficient. To illustrate how to perform the test for independence based on the Kendall coefficient in $R$ we consider the example about the hypermarkets. The $R$ code to test the hypothesis that the ratio between net profit and staff hours and the ratio between net profit and sale space of the hypermarkets are independent against the one-sided upper-tail alternative hypothesis that they are concordant is as follows:

```
> #net profit/staff hours, ranks
> x=c(1,2,3,6,4,10,8,15,9,5,7,17,13,12,11,14,16,18,19)
>
> #net profit/sale space, ranks
> y=c(1,3,2,6,5,7,11,9,13,14,8,12,4,17,16,15,18,10,19)
>
> cor.test(x,y,method="kendall",alternative="greater")
######################################################################
# Kendall's rank correlation tau
#
#data: x and y
#T = 131, p-value = 0.0005317
#alternative hypothesis: true tau is greater than 0
#sample estimates:
# tau
#0.5321637
######################################################################
```

The result shows that there is very strong evidence that the ratio between net profit and staff hours and the ratio between net profit and sale space are concordant because the $p$-value is very small. It should be cautioned that the dataset is not a random sample of observations because all hypermarkets of the Prisma chain in Finland have been considered, rather than a random sample of them or more generally a random sample from the population of all hypermarkets in Finland. For the purpose of illustration of the procedure in $R$ we treated the dataset as a random sample. To obtain the $p$-value $\Pr\{N_c \geq 131\} = \Pr\{\tau_b \geq 0.53216\}$ run

```
> cor.test(x,y,method="kendall",alternative="greater")$p.value
[1] 0.0005316611
```

and to obtain the test statistic $N_c$ run

```
> cor.test(x,y,method="kendall",alternative="greater")$statistic
  T
131
```

Note that $\binom{n}{2} = \frac{19 \cdot 18}{2} = 171$ therefore $N_d = 171 - N_c = 40$ and $\tau = \frac{131-40}{171} = 0.53216$. To test the null hypothesis of independence against the one-sided lower-tail alternative that $X$ and $Y$ are discordant just replace alternative="greater" with alternative="less". To test the null hypothesis against the two-sided alternative that $X$ and $Y$ are either concordant or discordant replace alternative="greater" with alternative="two.sided". It is important to note that for $n < 50$ and in the absence of ties $R$ computes exactly the $p$-value of the test. With $n \geq 50$ and/or in the presence of ties, the $N_c - N_d$ statistic (which is the numerator of $\tau_b$ and is one to one increasingly related to it) is standardized to 0 mean and 1 variance and an approximate $p$-value based on the asymptotic normality of this statistic is computed.

Reanalyzing the data using the Spearman test of independence we obtain very similar results because the $p$-value of the test, obtained by running

```
> cor.test(x,y,method="spearman",alternative="greater")$p.value
[1] 0.00119041
```

is 0.001. Conover (1999) underlined that there is no strong reason to prefer one test over the other one because generally both tests produce nearly identical results.

It is important to emphasize that the Kendall test, like the Spearman test, is a distribution-free test because the null distribution of the test statistic can be obtained without any assumptions on the distributions of the populations behind $X$ and $Y$ except requiring that they are continuous.

## 6.4    Tests for concordance

Rankings of a group of objects are very often considered in human resource management, education, marketing, politics and finance, when job applicants, new products, political parties, private or public services, firms and investments are ranked by executives, head hunters, experts, focus groups, investors or even automated algorithms (such as those considered by Marozzi, 2009, 2012a,c). An important question that naturally arises is whether the rankings given by a set of $p$ criteria (or judges) show any agreement or are more or less independent.

The most familiar measure for concordance is the Kendall $W$ coefficient (Kendall and Babington Smith, 1939) that has been applied to many different situations. For example, Grothe and Schmid (2011) used Kendall $W$ in finance to study associations of volatilities of asset returns. Association of asset returns is a useful indicator of asset portfolio diversification. Volatility of asset returns is a central aspect in the valuation of corresponding derivatives and then association of volatilities is a useful indicator for the derivatives market. Legendre (2005) used Kendall $W$ in ecology to search for species associations, that is groups of species that are found together. This is an important problem of community ecology because species associations may be used to predict environmental characteristics. Classical tests for concordance between several judges/criteria are the Friedman (1937) test and the $F$ test (Kendall and Babington Smith, 1939). Note that this problem is different from testing for evidence of agreement between two groups of judges (e.g., male and female judges; Schucany and Frawley, 1973; Hollander and Sethuraman, 1978; Vanbelle and Albert, 2009). Kraemer (1981) and Feigin and Alvo (1986) addressed the general case of two or more groups of judges.

Legendre (2005) showed via simulation that the Friedman test is too conservative and less powerful than its permutation version which always has a correct size. Unfortunately, the simulation study of Legendre was very limited because it considered neither the modeling of the copula of the underlying multivariate distribution nor the $F$ test. Kendall $W$ is a rank based correlation measure and therefore it is not affected by the marginal distributions of the underlying variables but only by the copula of the multivariate distribution (Grothe and Schmid, 2011). Marozzi (2013) greatly extended the simulation study of Legendre by modeling the copula of the underlying multivariate distribution as well as the $F$ test. His object was twofold: first to study the size and power of the $F$ test, and secondly to find out whether the conclusions drawn by Legendre (2005) are more general. It is shown that the Friedman test is too conservative and less powerful than both the $F$ test and the permutation test for concordance which always have a correct size and behave alike. Surprisingly, it is also shown that the power functions of the tests are not much affected by the type of copula.

Section 6.4.1 presents the Kendall–Babington Smith $F$ test and Section 6.4.2 presents a permutation test for concordance. An application to a very important financial problem is presented. More precisely, we consider a random sample of publicly traded USA firms and a set of valuation ratios. We would like to understand whether the valuation ratios are concordant or not.

## 6.4.1    The Kendall–Babington Smith test

Let $R_{ij}$ denote the rank of the $i$th object ($i = 1, \ldots, n$) given by the $j$th judge ($j = 1, \ldots, p$) and let

$$R_i = \sum_{j=1}^{p} R_{ij}$$

be the sum of the ranks for object $i$. These sums reflect the degree of concordance among the judges. On the one hand, when there is little or no concordance $R_1, \ldots, R_n$ are approximately equal. On the other hand, when there is perfect concordance the $p$ rankings are identical and $R_1, \ldots, R_n$ would be as different as possible. Therefore Kendall and Babington Smith (1939) considered the variance $\frac{1}{n} \sum_{i=1}^{n} (R_i - \overline{R})^2$ of the sums of ranks $R_i$ as a measure of concordance among the judges, where $\overline{R} = \frac{1}{n} \sum_{i=1}^{n} R_i$.

By dividing it by its maximum $\frac{p^2(n^3-n)}{12n}$, Kendall and Babington Smith (1939) obtained the so called Kendall coefficient of concordance

$$W = \frac{12 \sum_{i=1}^{n} (R_i - \overline{R})^2}{p^2(n^3 - n) - pT} = \frac{12 \sum_{i=1}^{n} R_i^2 - 3p^2 n(n + 1)^2}{p^2(n^3 - n) - pT}$$

where $T$ is a correction factor for tied ranks

$$T = \sum_{k=1}^{m} (t_k^3 - t_k)$$

and $t_k$ is the number of tied ranks in each of $m$ groups of ties. Of course $W \in [0, 1]$. It is interesting to note that $W$, in the absence of ties, can be expressed as a simple transformation of the mean $\overline{\rho}$ of the Spearman correlation coefficients among ranks of all pairs of judges or criteria

$$W = \frac{(p - 1)\overline{\rho} + 1}{p}.$$

When $p = 2$, $W$ is a linear transformation of the Spearman correlation coefficient $\rho$ between the ranks given by the two judges. If the ranks are perfectly discordant then $\rho = -1$ and $W = 0$; if the ranks are independent then $\rho = 0$ and $W = 0.5$; if the ranks are perfectly concordant then $\rho = 1$ and $W = 1$.

As emphasized by Marozzi (2013), the Kendall coefficient of concordance can be applied to data collected for estimating the same general property of the objects. This general property is a latent variable that cannot be directly observed. For example in finance, think about a set of financial ratios that assess profitability, efficiency, activity and liquidity of a set of firms (the objects). They are used by financial analysts or investors (the judges) to evaluate how sound would be an investment in equity or debt issued by a certain firm. The soundness of an investment is a latent variable because it cannot be directly measured.

Classical tests of the coefficient of concordance are the Friedman $C$ test and the Kendall–Babington Smith $F$ test. The null hypothesis $H_0$ is the independence of the rankings given by the judges. The alternative hypothesis $H_1$ is that at least one judge is concordant with at least one of the other judges. Note that the tests on the Kendall coefficient of concordance are one-sided because only positive associations between ranks are recognized. The Friedman (1937) $C$ statistic has been proposed to perform the two-way analysis of variance without replication by ranks and is related to $W$ by the equation

$$C = p(n-1)W.$$

Under the null hypothesis, the distribution of $C$ tends to that of $\chi^2$ with $n-1$ degrees of freedom as $p$ tends to infinity. Therefore the null hypothesis is rejected at the $\alpha$ nominal significance level if $C \geq \chi^2_{n-1,1-\alpha}$, where $\chi^2_{n-1,1-\alpha}$ denotes the $(1-\alpha)\%$ percentile of the $\chi^2$ distribution with $n-1$ degrees of freedom. Kendall and Babington Smith (1939) emphasized that the $\chi^2$ approximation is satisfactory for moderately large values of $p$, while for small values the fit near the tails is not likely to be very good because the approximation is affected by the disadvantage of representing a distribution of finite range by one of infinite range.

Kendall and Babington Smith (1939) proposed a different test by noting that the first four moments of the Pearson type I distribution

$$\frac{1}{B(a,b)} W^{a-1}(1-W)^{b-1}$$

with $a = (n-1)/2 - 1/p$ and $b = (p-1)a$, where $B$ denotes the beta function, are approximately those of $W$ if $p$ and $n$ are moderately large. It is most convenient to put

$$F = \frac{W(p-1)}{1-W}$$

so that under the null hypothesis, the $F$ statistic is asymptotically distributed as the $F$ distribution with $v_1 = n - 1 - 2/p$ and $v_2 = v_1(p-1)$ degrees of freedom. The null hypothesis is rejected at the $\alpha$ nominal significance level if $F \geq F_{v_1,v_2,1-\alpha}$, where $F_{v_1,v_2,1-\alpha}$ denotes the $(1-\alpha)\%$ percentile of the the $F$ distribution with $v_1$ and $v_2$ degrees of freedom. It is important to emphasize that the $F$ test is not as familiar as the $C$ test.

We use the Kendall–Babington Smith $F$ test to assess a very important financial problem: to understand whether a set of firm financial ratios is concordant or not. Financial ratios represent the key financial characteristics of the firm. Financial analysts, managers, lenders and academic researchers widely use financial ratios. Financial analysts use them to predict how well the securities of one company will perform relative to that of another one. Managers use them to know what divisions have performed well or to know when existing capacity will be exceeded. Lenders use them to predict if the borrower will be able to sustain interests and pay the principal. Common applications of financial ratios in academic researches include distress and failure prediction studies, trend analysis studies of individual company performance and cross-sectional studies comparing individual company ratios versus industry average ratios. Financial ratios are computed from company annual reports and required disclosures, in particular for publicly traded companies. Ratios measuring profitability, activity, efficiency and liquidity are considered (Barnes, 1987). Chen and Shimerda (1981) and Hossari and Rahman (2005) reviewed the literature finding 41 and 48 ratios, respectively, to be used in practice. In the literature, there is no clear indication as to which are the most or least important financial ratios and some authors call for circumspection in the use of financial ratios (Sudarsanam and Taffler, 1995). Marozzi and Santamaria (2010) and Marozzi (2012b) proposed a quick and simple method for selecting financial ratios according to their relevance in ranking a group of firms. Here we consider valuation ratios. In particular, we consider the following ratios, which are very popular (Damodaran, 2006):

- $X_1 = P/E = $ price to earnings ratio $= \frac{\text{market capitalization}}{\text{net income}}$. It represents the market capitalization of a firm as a multiple of its net income. It shows how much investors are willing to pay per unit of earnings. Firms trading at high $P/E$ are expected to show higher earnings growth in the future compared with firms with lower $P/E$, for this reason the former are more expensive than the latter. The fundamental determinants of the $P/E$ ratio are the expected growth rate in earnings per share, the payout and the risk.

- $X_2 = P/B = $ price to book equity ratio $= \frac{\text{market capitalization}}{\text{current book value of equity}}$. It represents the market capitalization of a firm as a multiple of its book value of equity. It shows how much investors are willing to pay per unit of book value of equity. Firms trading at high $P/B$ are expected to create in the future more value from their assets than those firms trading at lower $P/B$. The fundamental determinants of the $P/B$ ratio are the expected growth rate in earnings per share, the payout, the risk and the return on equity.

- $X_3 = P/S = $ price to sales ratio $= \frac{\text{market capitalization}}{\text{revenues}}$. It represents the market capitalization of a firm as a multiple of its revenues. Firms trading at high $P/S$ are expected to show higher revenues growth in the future compared with firms with lower $P/S$. The fundamental determinants of the $P/S$ ratio are the expected growth rate in earnings per share, the payout, the risk and the net margin.

- $X_4 = EV/EBITDA = $ enterprise value to EBITDA ratio $= \frac{\text{enterprise value}}{\text{EBITDA}}$, where the enterprise value is the market value of debt and equity of a firm net of cash and EBITDA stands for earnings before interest, taxes, depreciation and amortization. $X_4$ represents the enterprise value of a firm as a multiple of its EBITDA. It shows how much a potential bidder is willing to pay to acquire the firm, including its debt position, per unit of EBITDA. Firms trading at high $EV/EBITDA$ are expected to improve their EBITDA in the future more than firms trading at lower $EV/EBITDA$. The fundamental determinants of the $EV/EBITDA$ ratio are the expected growth rate in earnings per share, the reinvestment rate, the risk, the return on invested capital and the tax rate.

- $X_5 = EV/C = $ enterprise value to capital ratio $= \frac{\text{enterprise value}}{\text{current invested capital}}$. It represents the enterprise value of a firm as a multiple of its invested capital. It shows how much a potential bidder is willing to pay to acquire the firm, including its debt position, per unit of invested capital. Firms trading at high $EV/C$ are expected to improve their investing projects in the future (i.e., to be wealth creating firms) more than firms trading at lower $EV/C$. The fundamental determinants of the $EV/C$ ratio are the expected growth rate in earnings per share, the reinvestment rate, the risk and the return of capital.

- $X_6 = EV/S$ = enterprise value to sales ratio $= \frac{\text{enterprise value}}{\text{revenues}}$. It represents the enterprise value of a firm as a multiple of its revenues. It shows how much a potential bidder is willing to pay to acquire the firm, including its debt position, per unit of revenues. Firms trading at high $EV/S$ are expected to increase their revenues in the future more than firms trading at lower $EV/S$. The fundamental determinants of the $EV/S$ ratio are the expected growth rate in earnings per share, the reinvestment rate, the risk and the operating margin.

Damodaran (2006) emphasizes that there have been relatively few studies that compare the efficacy of the financial ratios and notes that the usage of them varies widely across industries with the $EV/EBITDA$ ratio commonly used for valuing heavy infrastructure firms (such as telecommunication ones) and the $P/B$ ratio for financial service firms. $P/E$ and $EV/EBITDA$ ratios are the most frequently used by the research arms of investment banks. Of course if $X_{ij} > X_{hj}$ then firm $i$ is valued higher than firm $h$ according to valuation ratio $X_j$.

We consider a random sample of 30 firms taken from publicly traded USA firms and their valuation ratios. The aim of our analysis is to understand whether the $p = 6$ financial ratios (the criteria) of the $n = 30$ firms (the objects) are concordant or not. In other words, the question is whether the rankings of the firms according to the six valuation ratios listed above show any agreement or are more or less independent. To answer this question we have to test the null hypothesis that the rankings of the firms are independent against the one-sided alternative that at least one valuation ratio is concordant with at least one of the other valuation ratios. To test this system of hypotheses we use the $F$ test for concordance.

The additional $R$ package vegan allows the Kendall $W$ coefficient of concordance to be computed and the $F$ test for concordance to be performed very easily in $R$. The package can be freely downloaded from
    http://cran.r-project.org/web/packages/vegan/index.html.
To run vegan it is necessary to load the permute additional package which can be freely downloaded from
    http://cran.r-project.org/web/packages/permute/index.html.
To perform our analysis, first we load the dataset and the additional packages

```
> library(permute)
> library(vegan)
> data=read.table("C:\\Data\\finance.csv",dec = ",",sep = ";",
    header=TRUE)
```

Secondly, we compute the Kendall $W$ coefficient of concordance

```
> kendall.global(data)$Concordance_analysis[1]
[1] 0.4688643
```

The result shows that the degree of concordance among the valuation ratios is about 47% of the maximum possible degree of concordance, that is when the rankings of the firms according to the various financial ratios are identical. Finally, to test whether the degree of concordance is significantly greater than zero we run

```
> #F statistic
> kendall.global(data)$Concordance_analysis[2]
[1] 4.413791
> #F test p-value
> kendall.global(data)$Concordance_analysis[3]
[1] 1.345288e-09
```

The $p$-value shows that there is very strong evidence in favor of the alternative hypothesis that the valuation ratios are concordant or more precisely that at least one valuation ratio is concordant with at least one of the other valuation ratios.

Marozzi (2013) emphasized that as most parametric tests, the reference null distributions of the $C$ test and the $F$ test are only known asymptotically, therefore the tests are guaranteed to have a correct size only asymptotically. This very important point is addressed in Section 6.4.2 where the permutation test for concordance, which is guaranteed to have a correct size also for small sample sizes (provided that the object rankings are exchangeable under the null hypothesis), is presented.

## 6.4.2    A permutation test for concordance

In Section 6.4.1 we presented the Kendall–Babington Smith $F$ test which is guaranteed to have a correct size only asymptotically because the reference null distribution of the corresponding test statistic is only known asymptotically. In the present section we present a test that is guaranteed to have a correct size also for small sample sizes. It is the permutation test on $W$ provided that the object rankings are exchangeable under the null hypothesis. Since the null hypothesis is the independence of the object rankings given by the judges, exchangeability holds and permuting all rank vectors independently of one another is justified. Note that

$$S = \sum_{i=1}^{n} R_i^2,$$

$W$, $C$, and $F$ statistics are permutationally equivalent and then it is not important which one is used as the pivotal statistic. Let $\boldsymbol{R}_j = (R_{1j}, \ldots, R_{nj})$ be the vector of object ranks given by judge $j$. The steps for performing the permutation $S$ test of concordance are:

1. Compute the observed value of the test statistic $S^o = S$.

2. Randomly permute (independently of one another) $R_j$ for $j = 1, \ldots, p$ and compute $S_1^* = \sum_{i=1}^{n}(R_i^*)^2$, where $R_i^* = \sum_{j=1}^{p} R_{ij}^*$ and $R_{ij}^*$ is the permuted rank of object $i$ given by judge $j$.

3. Independently repeat step 2 for $B$ times.

4. Compute the permutation test $p$-value as

$$\lambda = \frac{1}{B} \sum_{b=1}^{B} \mathbb{I}\left(S_b^* \geq S^o\right),$$

   where $\mathbb{I}(\cdot)$ denotes the indicator function.

5. Reject the null hypothesis at the $\alpha$ nominal level of significance if $\lambda \leq \alpha$, otherwise accept it.

It is important to emphasize that all the tests on the Kendall coefficient of concordance are one-sided because only positive associations between ranks are recognized. It is worth noting that Legendre (2005) computes the $p$-value of the test as

$$\lambda' = \frac{1}{B+1}\left[\sum_{b=1}^{B}\mathbb{I}\left(S_b^* \geq S^o\right) + 1\right] = \frac{1}{B+1}\left[\sum_{b=0}^{B}\mathbb{I}\left(S_b^* \geq S^o\right)\right].$$

As shown by Pesarin and Salmaso (2010, p. 68–69), this is a biased estimator of the true $p$-value, whereas $\lambda$ is unbiased. However, using $\lambda$ or $\lambda'$ is equivalent in practice unless $B$ is very small.

Marozzi (2013) used a copula based approach to study the size and power of the Friedman $C$ test, the Kendall–Babington Smith $F$ test and the permutation $S$ test. He showed that the Friedman test is too conservative and less powerful than both the $F$ test and the permutation test for concordance which always have a correct size and behave alike. Surprisingly, it is also shown that the power functions of the tests are not much affected by the type of copula.

To illustrate how the permutation test for concordance can be performed in $R$ we analyze again the financial dataset about valuation ratios described in Section 6.4.1. We have $n = 30$ firms (the objects) and $p = 6$ valuation ratios (the criteria) and we would like to test the null hypothesis that the rankings of the firms according to the valuation ratios are independent against the one-sided alternative that at least one valuation ratio is concordant with at least one of the other valuation ratios. To perform the permutation test for concordance in $R$ run the following code.

```
> library(permute)
> library(vegan)
```

```
> data=read.table("C:\\Documents\\finance.csv",dec = ",",sep = ";",
  header=TRUE)
> #permutation test p-value
> kendall.global(data,nperm=10,000)$Concordance_analysis[5]
[1] 9.999e-05
```

where vegan is an additional $R$ package that, among many other things, computes the Kendall coefficient of concordance and performs hypothesis testing on it, permute is an additional $R$ package required by vegan (see Section 6.4.1 for how to download these additional packages) and where nperm is the number $B$ of permutations used for estimating the $p$-value of the test. The default for nperm is 1000. Note that since we set a large number for $B$ then $\lambda'$ should not be corrected as $\lambda$. However, in case you should, or want to, correct $\lambda'$ just compute

$$\lambda = \frac{\lambda'(B+1)-1}{B}.$$

The result of the permutation $S$ test is consistent with the result of the Kendall–Babington Smith $F$ test (see Section 6.4.1) in concluding that there is very strong evidence in favor of the alternative hypothesis that the valuation ratios are concordant or more precisely that at least one valuation ratio is concordant with at least one of the other valuation ratios.

# References

Barnes, P. (1987) The analysis and use of financial ratios: a review article. Journal of Business Finance and Accounting, Winter, 449–461.

Best, D.J. and Roberts, D.E. (1975) Algorithm AS 89: the upper tail probabilities of Spearman's rho. Applied Statistics, 24, 377–379.

Chen, K.H. and Shimerda, T.A. (1981) An empirical analysis of useful financial ratios. Financial Management, 10, 51–60.

Conover, W.J. (1999) Practical Nonparametric Statistics, 3rd edn. John Wiley & Sons, Ltd.

Damodaran, A. (2006) Damodaran on Valuation, 2nd edn. John Wiley & Sons, Ltd.

Feigin, P.D. and Alvo, M. (1986) Intergroup diversity and concordance for ranking data: an approach via metrics for permutations. The Annals of Statistics, 14, 691–707.

Friedman, M. (1937) The use of ranks to avoid the assumption of normality implicit in the analysis of variance. Journal of the American Statistical Association, 32, 675–701.

Grothe, O. and Schmid, F. (2011) Kendall's W reconsidered. Communications in Statistics – Simulation and Computation, 40, 285–305.

Hollander, M. and Sethuraman, J. (1978) Testing for agreement between two groups of judges. Biometrika, 65, 403–411.

Hossari, G. and Rahman, S. (2005) A comprehensive formal ranking of the popularity of financial ratios in multivariate modeling of corporate collapse. Journal of American Academy of Business, 6, 321–327.

Kendall, M.G. and Babington Smith, B. (1939) The problem of m rankings. The Annals of Mathematical Statistics, 10, 275–287.

Korhonen, P. and Siljamäki, A. (1998) Ordinal principal component analysis theory and an application. Computational Statistics & Data Analysis, 26, 411–424.

Kraemer, H.C. (1981) Intergroup concordance: definition and estimation. Biometrika, 68, 641–646.

Legendre, P. (2005) Species associations: the Kendall coefficient of concordance revisited. Journal of Agricultural, Biological, and Environmental Statistics, 10, 226–245.

Marozzi, M. (2009) A composite indicator dimension reduction procedure with application to university student satisfaction. Statistica Neerlandica, 63, 258–268.

Marozzi, M. (2012a) Tertiary student satisfaction with socialization: a statistical assessment. Quality and Quantity, 46, 1271–1278.

Marozzi, M. (2012b) Composite indicators: a sectorial perspective. In: Perna, C. and Sibillo, M. (eds) Mathematical and Statistical Methods for Actuarial Sciences and Finance. Springer, pp. 287–294.

Marozzi, M. (2012c) Construction, dimension reduction and uncertainty analysis of an index of trust in public institutions. Quality and Quantity, DOI 10.1007/s11135-012-9815-z.

Marozzi, M. (2013) Testing for concordance between several criteria. Journal of Statistical Computation and Simulation, DOI:10.1080/00949655.2013.766189.

Marozzi, M. and Santamaria, L. (2010) A dimension reduction procedure for corporate finance indicators. In: Corazza, M. and Pizzi, C. (eds) Mathematical and Statistical Methods for Actuarial Sciences and Finance. Springer, pp. 205–213.

Pesarin, F. and Salmaso, L. (2010) Permutation Tests for Complex Data. John Wiley & Sons, Ltd.

Reimann, B.C. (1974) Dimensions of structure in effective organizations: some empirical evidence. Academy of Management Journal, 17, 693–708.

Schucany, W.R. and Frawley, W.H. (1973) A rank test for two group concordance. Psychometrika, 38, 249–258.

Sudarsanam, P.S. and Taffler, R.J. (1995) Financial ratio proportionality and inter-temporal stability: an empirical analysis. Journal of Banking & Finance, 19, 45–60.

Vanbelle, S. and Albert, A. (2009) Agreement between two independent groups of raters. Psychometrika, 74, 477–491.

# 7

# Tests for heterogeneity

## 7.1   Introduction

When dealing with numerical variables, some of the most interesting and useful inferential problems concern estimating and testing for variances. Since the variance represents the population functional that measures the variability, that is the tendency of the variable of assuming data with a given degree of dispersion around the mean, these inferential problems are very important in many fields of application. Examples include: in statistical quality control and in performance analysis, reduction of waste and improvement of performance are related to lowering of variances; the efficiency of any estimator can be measured through its variance; and tests for the equality of variances are important for choosing the appropriate test statistic in the two-sample $t$ test for mean comparisons, see Section 2.2.

In the presence of categorical data, unless it is possible and sensible to transform the original variables assigning suitable scores to the modalities, mean and variance, as a measure of central tendency and variability of a distribution, respectively, cannot be computed. In many problems, for example in the presence of nominal variables, score transformation is nonsense. Let us consider that several real phenomena may be represented only by nominal categorical variables and many others by ordinal variables, the score transformation is subjective and questionable because it may change the original information provided by data (e.g., in opinion polls, performance qualitative assessments, psycho aptitude tests, etc.). In these cases score transformation should not be used. Some indexes like median and mode can be considered for measuring the central tendency of the distribution. Instead of variability the notion of heterogeneity may be used for categorical variables. The notion of heterogeneity is independent of the ordinal or nominal nature of the variables because it does not consider the possible ordering of data. Furthermore the notion of statistical heterogeneity can be applied in

*Nonparametric Hypothesis Testing: Rank and Permutation Methods with Applications in R,*
First Edition. Stefano Bonnini, Livio Corain, Marco Marozzi and Luigi Salmaso.
© 2014 John Wiley & Sons, Ltd. Published 2014 by John Wiley & Sons, Ltd.
Companion website: http://www.wiley.com/go/hypothesis_testing

Economics, Genetics, Biology, Engineering and other sciences for measuring several different phenomena such as external efficacy of academic courses (heterogeneity of job opportunities), market segmentation, genetic differentiation, biodiversity, spatial clustering, etc. For example, an interesting problem of genetics is related to the comparison of two populations in order to verify which of them has a wider genetic differentiation, that is a wider heterogeneity from the point of view of the phenotypical combinations of certain genetic factors (Corrain *et al.*, 1977). Ecological diversity is another typical application of the index of heterogeneity (Pielou, 1975, 1977; Patil and Taillie, 1982). The degrees of heterogeneity of postdoctoral judgments in two surveys about PhD satisfaction with regard to the adequacy of the PhD education for the work carried out were compared in Arboretti Giancristofaro *et al.* (2009b).

Let us assume that $C$ compared populations $(C \geq 2)$ may take $K$ categories and let $f_{jk}$ denote the absolute frequency of the $j$th sample for the $k$th category $(j = 1, \ldots, C; k = 1, \ldots, K)$. The typical data of the problem consist of the $C \times K$ contingency table of the observed frequencies $[f_{jk}]$.

## 7.2    Statistical heterogeneity

In descriptive statistics a variable $X$ is said to be *minimally heterogeneous* or equivalently *maximally homogeneous* when all the $n$ observed statistical units present the same modality, and then its distribution is degenerate. Conversely $X$ is *maximally heterogeneous (minimally homogeneous)* when all the modalities/categories are observed with the same frequency, that is the distribution is uniform over the set of modalities. Thus, heterogeneity is strictly related to the concentration of frequencies.

From the inferential point of view let us consider a categorical response variable $X$ and let us suppose that it takes categories in $(A_1, \ldots, A_K)$, with unobserved probability distribution $\Pr\{X = A_k\} = \pi_k, k = 1, \ldots, K$. The following properties must be satisfied by any index $\eta$ measuring the degree of heterogeneity:

1. It reaches its minimum when the distribution is degenerate, that is when there is an integer $r \in (1, \ldots, K)$ such that $\pi_r = 1$ and $\pi_k = 0, \forall k \neq r$.

2. It assumes increasingly greater values when moving away from the degenerate towards the uniform distribution.

3. It reaches its maximum when the distribution is uniform, that is $\pi_k = 1/K$, $\forall k \in (1, \ldots, K)$.

Various indexes satisfy the three properties and can be used to measure the degree of heterogeneity. For a detailed discussion on heterogeneity in descriptive statistics see Piccolo (2000). The following are some of the most commonly used indexes:

- Gini index (Gini, 1912): $\eta_G = \sum_{k=1}^{K} \pi_k(1 - \pi_k) = 1 - \sum_{k=1}^{K} \pi_k^2$

- Shannon entropy index (Shannon, 1948): $\eta_S = - \sum_{k=1}^{K} \pi_k \log(\pi_k)$

- Rényi entropy of order $\delta$ (Rényi, 1966): $\eta_{R_\delta} = \frac{1}{1-\delta} \log \sum_{k=1}^{K} \pi_k^\delta$, where $\log(\cdot)$ is the natural logarithm and assuming that $0 \cdot \log(0) = 0$. It can be trivially proved that $\eta_{R_1} = \eta_S$ and $\eta_{R_2} = -\log[1 - \eta_G]$. In the family of indexes defined by Rényi we will consider $\eta_{R_3} = -\frac{1}{2} \log \sum_{k=1}^{K} \pi_k^3$ and $\eta_{R_\infty} = -\lim_{\delta \to \infty} \eta_{R_\delta} = -\log[\max_k(\pi_k)]$. Among the most interesting proposals about indexes for measuring heterogeneity we include also Leti (1965) and Frosini (1981).

## 7.3   Dominance in heterogeneity

Given two populations $P_1$ and $P_2$, let us denote with $Het(P_j)$ the heterogeneity of $P_j$, and with $\eta_j$ an index (Gini's, Shannon's, Rényi's, or others) for measuring $Het(P_j)$, $j = 1, 2$. Let us suppose an interest in testing the hypotheses

$$H_0 : Het(P_1) = Het(P_2)$$

against the alternative

$$H_1 : Het(P_1) > Het(P_2).$$

As a matter of fact, if probabilities $\{\pi_{jk}, k = 1, \ldots, K, j = 1, 2\}$ were known, they could be arranged in non-increasing order $\pi_{j(1)} \geq \ldots \geq \pi_{j(K)}$, according to the Pareto diagram rule. The null hypothesis could be written as

$$H_0 : \{Het(P_1) = Het(P_2)\} = \{\pi_{1(k)} = \pi_{2(k)}, k = 1, \ldots, K\}.$$

Under the null hypothesis, exchangeability holds and the permutation testing principle is applicable exactly. The alternative hypothesis of the problem may be defined as

$$H_1 : \{Het(P_1) > Het(P_2)\} = \left\{ \sum_{s=1}^{k} \pi_{1(s)} \leq \sum_{s=1}^{k} \pi_{2(s)}, k = 1, \ldots, K \right\}$$

and the strict inequality holds for at least one $k = 1, \ldots, K$.

Let us consider two independent samples with i.i.d. observations $X_j = \{X_{ji}, i = 1, \ldots, n_j; n_j > 2\}$, $j = 1, 2$. Observed data can be displayed in a $2 \times K$ contingency table with absolute frequencies $\{f_{jk} = \sum_{i \leq n_j} \mathbb{I}(X_{ji} = A_k), k = 1, \ldots, K, j = 1, 2\}$. Marginal frequency of the $k$th column (category) is indicated by $f_{\bullet k} = f_{1k} + f_{2k}, k = 1, \ldots, K$; and marginal frequency of the $j$th row is the sample size $n_j, j = 1, 2$.

The observed dataset may be also represented by $X = \{X(i), i = 1, \ldots, n; n_1, n_2\}$. A permutation of the rows of the data implies that some units of sample 1 are reassigned to sample 2 and vice versa. Thus the contingency table of a permuted dataset $\mathbf{X}^*$ (the permuted table) has the same marginal frequencies of the observed contingency table. In other words, marginal frequencies are permutation invariant, that is $f_{\bullet k} = f_{1k} + f_{2k} = f_{1k}^* + f_{2k}^* = f_{\bullet k}^*$, $k = 1, \ldots, K$, where $f_{jk}^*$, $j = 1, 2$, are the frequencies of the permuted table.

It is worth observing that if the exchangeability condition is satisfied, in univariate two-sample tests the set of marginal frequencies $(n_1, n_2, f_{\bullet 1}, \ldots, f_{\bullet K})$, the dataset $X$, and any of its permutations $X^*$, are equivalent sets of sufficient statistics (Pesarin and Salmaso, 2010). If exchangeability holds only asymptotically, only approximate solutions are possible. The permutation tests for heterogeneity fall into this category of procedures.

Hence heterogeneity of a population may be defined in terms of concentration of probabilities. As a consequence heterogeneity comparisons may be performed comparing the cumulative probabilities of the Pareto diagram. In this sense, the problem is in some ways similar to that of stochastic dominance, for ordered categorical variables, even if the ordering criterion is based on the probabilities instead of the ordered categories. For problems of stochastic dominance in the literature there is quite a long list of exact and approximate solutions. We mention those of Hirotsu (1986), Lumley (1996), Loughin and Scherer (1998), Nettleton and Banerjee (2001), Han *et al.* (2004), Loughin (2004) and Agresti and Klingerberg (2005). Most of the methodological solutions proposed for the univariate case are based on the restricted maximum likelihood ratio test (Cohen *et al.*, 2000; Silvapulle and Sen, 2005). In these solutions, under the null and alternative hypothesis the test statistics asymptotically follow mixtures of chi-square distributions with weights essentially dependent on the unknown population distribution. Pesarin (1994, 2001), Brunner and Munzel (2000), Troendle (2002) and Pesarin and Salmaso (2006) propose nonparametric methods.

Unfortunately, ordered parameters $\pi_{j(k)}$, $k = 1, \ldots, K, j = 1, 2$, are unknown and can only be estimated using the observed ordered frequencies: $\hat{\pi}_{j(k)} = f_{j(k)}/n_j$. It implies that the ordering within each population, estimated through relative frequencies within each sample, is a *data-driven ordering* and so it may differ from the true one and it presents sampling variability. Hence exchangeability under $H_0$ is not exact but only approximated. Data-driven and true ordering are equal with probability one only asymptotically and, from the well-known Glivenko–Cantelli theorem (Shorack and Wellner, 1986), we can say that exchangeability of data with respect to samples is asymptotically attained under the null hypothesis.

To solve the testing problem, a reasonable test statistic could be a function of the sampling estimates of the indexes described in Section 7.2. Specifically, considering the sampling index $\hat{\eta}_j = \eta(f_{j1}/n_j, \ldots, f_{jK}/n_j)$ as estimate of the index $\eta_j = \eta(\pi_{j1}, \ldots, \pi_{jK}), j = 1, 2$, a possible test statistic could be $T_\eta = \hat{\eta}_1 - \hat{\eta}_2$, where $\eta$ is any index of heterogeneity (Gini index, Shannon index, Rényi index, or others). According to the chosen index of heterogeneity, for example we can have the following alternative test statistics:

$$T_G = \sum_{k=1}^{K} [(f_{2k}/n_2)^2 - (f_{1k}/n_1)^2],$$

$$T_S = \sum_{k=1}^{K} [(f_{2k}/n_2) \log(f_{2k}/n_2) - (f_{1k}/n_1) \log(f_{1k}/n_1)],$$

$$T_{R_3} = \frac{1}{2}\left[\log\sum_{k=1}^{K}(f_{2k}/n_2)^3 - \log\sum_{k=1}^{K}(f_{1k}/n_1)^3\right],$$

$$T_{R_\infty} = \log\max_{k=1,\ldots,K}(f_{2k}/n_2) - \log\max_{k=1,\ldots,K}(f_{1k}/n_1),$$

$H_0$ should be rejected in favor of $H_1$ for large values of $T_\eta$. It is worth noting that all the defined indexes of heterogeneity are order invariant, that is $\eta_j = \eta(\pi_{j1}, \ldots, \pi_{jK}) = \eta(\pi_{j(1)}, \ldots, \pi_{j(K)})$, $j = 1, 2$ and similarly $\hat{\eta}_j = \eta(\hat{\pi}_{j1}, \ldots, \hat{\pi}_{jK}) = \eta(\hat{\pi}_{j(1)}, \ldots, \hat{\pi}_{j(K)})$, $j = 1, 2$. In the $2 \times K$ contingency table of ordered frequencies (ordered table) $\{f_{j(k)};\ k = 1, \ldots, K; j = 1, 2\}$, where $f_{j(1)} \geq \ldots \geq f_{j(k)}$, the $k$th column (ordered category) for sample 1 may correspond to an original category different from the one of the $k$th column in sample 2. Furthermore, in the case of ties in frequencies, we can arbitrarily choose their order since this does not affect the permutation analysis.

After the computation of the observed ordered table for both of the samples, it is possible to proceed with the application of a testing procedure for stochastic dominance on ordered categorical variables:

1. The observed value of the test statistics $T_\eta^o$ is calculated.

2. $B$ independent permutations of the dataset are performed and for each permutation, a permuted table $\{f_{j(k)}^*;\ k = 1, \ldots, K; j = 1, 2\}$ is obtained and the corresponding permuted value of the test statistic $T_{\eta(b)}^*$, $b = 1, \ldots, B$ is calculated.

3. Finally, the $p$-value $\lambda_\eta = \sum_{b=1}^{B}\mathbb{I}\left(T_{\eta(b)}^* \geq T_\eta^o\right)/B$ can be computed and compared with the significance level $\alpha$ as usual: $H_0$ is rejected in favor of $H_1$ if $\lambda_\eta \leq \alpha$, otherwise it cannot be rejected.

Using simulation studies Arboretti Giancristofaro et al. (2009a) prove that the testing procedure is well approximated under $H_0$ and that the tests based on the statistics $T_G$ and $T_S$ present a very similar power behavior and they are more powerful than $T_{R_3}$ and $T_{R_\infty}$.

## 7.3.1   Geographical heterogeneity

In 2012 the Faculty of Economics at the University of Ferrara (Italy) performed a survey to analyze the geographic place of residence of the students enrolled in the first year of the second level degree course 'Economics, Markets and Management' (EMM). Obviously most of the students live in Ferrara, the town where the university is located, but traditionally the geographic origin of the students of the second level degree is quite heterogeneous mainly for two reasons: (1) many students come from several towns and regions near Ferrara or even far away from it; (2) after the achievement of the first level degree, many students living in Ferrara prefer to move to other towns and enrol in a second level degree course at another university, making the geographic origin of the students of EMM more heterogeneous.

During the academic year 2010/11, that is from November 2010 to October 2011, some initiatives based on a *communication strategy* addressed to the first level degree students of the Faculty were instituted to improve the attractiveness of the EMM course among the students of Ferrara. Advertising, information about the degree program and career opportunities for graduate, orientation meetings and other similar actions were carried out. In the academic year 2011/12 the number of students from Ferrara enrolled in the first year of the EMM course was much greater (+33%). Hence the *communication strategy* was successful.

To study the changes in the geographic attractiveness of the course induced by the *communication strategy*, that is to test if with the growth of the number of students from Ferrara there was a concomitant significant reduction of the heterogeneity of the place of residence, a test for heterogeneity comparisons was performed. The response is therefore the nominal categorical variable representing the geographic origin and the modalities correspond to the places of residence.

To test the consequence of the *communication strategy* adopted by the Faculty of Economics, in terms of heterogeneity of the geographical origin of the students of the EMM course, a similar test can be applied, considering $P_1$ as the residence of the set of students enrolled in the course in the first year in 2010/11, that is before the application of the strategy, and $P_2$ as the residence of the set of students enrolled in the course in the first year in 2011/12, that is after the application of the strategy.

The problem described here concerns a typical example of geographical heterogeneity. If the goal of the problem consists of comparing the heterogeneity of the place of residence of the students in the academic year 2011/12, that is after the *communication strategy*, $Het(Residence_{2011})$, with the heterogeneity of the place of residence in the previous academic year, that is before the *communication strategy*, $Het(Residence_{2010})$, the null hypothesis of the problem is $H_0 : Het(Residence_{2010}) = Het(Residence_{2011})$ and the alternative hypothesis is $H_1 : Het(Residence_{2010}) > Het(Residence_{2011})$.

Table 7.1 shows the distributions of students according to the place of residence in the two considered academic years. It is evident that mostly the students come from the city of Ferrara. The communication strategy was successful because the percentage of students of Ferrara moved from 35 (45/130) to 40% (60/149). The absolute frequencies related to the other places did not change much hence, from a descriptive point of view, the set of students after the strategy became more homogeneous (less heterogeneous) in terms of geographical origin.

Table 7.2 indicates that when comparing the sampling indexes of the two groups the differences are not so evident. In particular, using the indexes of Gini and Shannon the values computed for the two samples are almost equal.

*R* codes to calculate sampling indexes and to perform permutation tests for heterogeneity comparisons are included in the file `heterogeneity.test.R`. Hence to run all the procedures and functions related to statistical analysis for heterogeneity comparisons you should call the source code:

```
> source("heterogeneity.test.R")
```

Table 7.1   Contingency table of the place of residence of the students enrolled in the first year of the EMM course (University of Ferrara) in the academic years 2010/11 and 2011/12.

| | Academic year | | |
| Place of residence | 2010/11 | 2011/12 | Total |
|---|---|---|---|
| Ferrara | 45 | 60 | 105 |
| Bologna | 3 | 4 | 7 |
| Ravenna | 6 | 5 | 11 |
| Rimini | 2 | 3 | 5 |
| Rovigo | 30 | 27 | 57 |
| Padova | 21 | 18 | 39 |
| Mantova | 3 | 5 | 8 |
| Puglia | 6 | 11 | 17 |
| Others | 14 | 16 | 30 |
| Total | 130 | 149 | 279 |

The Gini index for the heterogeneity of the residence of students in the first of the two compared academic years can be computed as follows:

```
> f1=c(45,3,6,2,30,21,3,6,14)
> G.index.1=het(f1,ind="G")
```

Function het has two arguments: the first corresponds to the vector of observed frequencies of the categorical variable, and the second denotes the index of heterogeneity to be computed. Hence the Gini index for the second academic year is obtained in a similar way:

```
> f2=c(60,4,5,3,27,18,5,11,16)
> G.index.2=het(f2,ind="G")
```

Table 7.2   Indexes of heterogeneity of the place of residence of the students enrolled in the first year of the EMM course (University of Ferrara) in the academic years 2010/11 and 2011/12.

| | Academic year | |
| Index | 2010/11 | 2011/12 |
|---|---|---|
| Gini | 0.784 | 0.770 |
| Shannon | 1.762 | 1.767 |
| Rényi(3) | 1.411 | 1.297 |
| Rényi($\infty$) | 1.061 | 0.910 |

The second argument `ind` can take the following values: `"G"` for calculating the Gini index; `"S"` to obtain the Shannon entropy; `"R3"` or `"Rinf"` for the computation of the Rényi index of order 3 or infinity, respectively.

To test the dominance in heterogeneity for the given problem you can use the `het.2sample.test` function as follows:

```
> f1=c(45,3,6,2,30,21,3,6,14)
> f2=c(60,4,5,3,27,18,5,11,16)
> F=cbind(f1,f2)
> het.2sample.test(F,alternative="greater",nperm=10 000)
```

The first input of the function is the contingency table F, defined as a $K \times 2$ table, where the first column is the vector of observed frequencies for the first sample `f1` and the second column is the vector of observed frequencies for the second sample `f2`. The second argument, `alternative`, indicates the alternative hypothesis of the test. If you wish to test the alternative hypothesis that population 1 has greater heterogeneity than population 2, such as in the considered example, then the correct choice is `"greater"`; if you wish to test the opposite one-sided alternative hypothesis, the correct choice is `"less"`. The third parameter needed by the function, `nperm`, is the number of independent permutations for estimating the permutation distribution of the test statistics and calculating the $p$-values. In this case 10000 permutations have been considered. A higher number of permutations guarantees the accuracy and robustness of the test but it also implies a lower efficiency in terms of required computing time to perform the procedure.

The final output is:

```
###################################################################
# Two-sample permutation test for heterogeneity comparisons:
# Alternative: greater
#
# Observed value of Gini's test statistic: 0.01361324
# P-value of Gini's test: 0.3236853
#
# Observed value of Shannon's test statistic: -0.00453116
# P-value of Shannon's test: 0.5011498
#
# Observed value of Renyi-3's test statistic: 0.1144874
# P-value of Renyi-3's test: 0.2328034
#
# Observed value of Renyi-infinity's test statistic: 0.1512702
# P-value of Renyi-infinity's test: 0.1645171
#
###################################################################
```

First of all the tested alternative hypothesis is shown. Then, for each of the four measures of heterogeneity, the observed value of the test statistic (difference between the sampling index of sample 1 and that of sample 2) and the corresponding $p$-value are given. In this case, even considering a high significance level, for example $\alpha = 0.10$, all the procedures consistently do not reject the null hypothesis because all the $p$-values are greater than $\alpha$. The minimum $p$-value (0.165) is that of Rényi's infinity test; the maximum $p$-value (0.501) is that of the procedure based on Shannon's index.

## 7.3.2  Market segmentation

For a company the evaluation of the heterogeneity of its customers is important for a suitable market segmentation. Market segmentation is fundamental for the application of product/service differentiation strategies. Market segmentation and corresponding differentiation strategies can give an important commercial advantage, but may be several theoretically 'ideal' market segments and every company could develop different ways of defining market segments. Hence the presence of significantly different segments of customers should be tested with sampling surveys, taking into account qualitative variables that could be important in the definition of segments. The analysis of statistical heterogeneity is strictly related to the segmentation analysis because high levels of heterogeneity in a group of customers indicates the presence of different segments.

The categorical variable under examination could be related to specific customs and habits of the customers or it could be the level of education, pastimes and lifestyle or any other socio-economic variable which is of interest to the company, considering also the specific type of product or service produced and/or delivered by the organization. Obviously the comparison of the heterogeneity of two distinct client groups, from the point of view of geographical areas, distribution channels, type of products or services, etc., could be useful to the company as a starting point for a segmentation and differentiation strategy.

In summer 2010 the Tourist Association of Sesto, a place in the mountains near Bolzano in the north of Italy, and the Natural Park of Dolomites of Sesto promoted the statistical survey 'Sesto Nature Survey' in order to know the satisfaction of visitors with regard to the facilities of the district of Sesto Dolomites/South Tyrol. In order to improve services and the protection of territory and to make holidays in the area more enjoyable, a sample of tourists was asked to complete a questionnaire about satisfaction, habits and preferences. One of the questions was related to the type of accommodation used during their stay in Sesto. One of the objectives was to analyze the ability to satisfy the demand for accommodation for holidaymakers, distinguishing between new and returning visitors. With this aim the hypothesis that the choice regarding the accommodation by tourists who had stayed previously in Sesto is less heterogeneous, mainly being oriented towards hotels, than that of tourists who were on their first visit to Sesto, could be of interest.

The categorical response variable *Accommodation* can take five modalities: *Camping*, *Hotel*, *Bed & Breakfast*, *Farm* and *House/Flat*. The factor useful to define the compared groups is *First visit to Sesto*, and it can take two levels: *no* and *yes*. Table 7.3 shows the observed frequencies.

The favorite accommodation is Hotel for both types of tourists. More than 80% of the returning visitors choose this accommodation while the percentage for new visitors is less than 70%. An interesting proportion of them prefer to stay in a

Table 7.3    Contingency table of the type of accommodation used by tourists in the 'Sesto Nature Survey' in 2010, with the distinction between tourists on their first visit or not.

| Accommodation | First visit | | Total |
| --- | --- | --- | --- |
| | No | Yes | |
| Camping | 1 | 4 | 5 |
| Hotel | 45 | 97 | 142 |
| Bed & Breakfast | 2 | 10 | 12 |
| Farm | 2 | 4 | 6 |
| House/Flat | 4 | 26 | 30 |
| Total | 54 | 141 | 195 |

House/Flat (18%) or in a Bed & Breakfast (7%). Note that for the statistical analysis with $R$, instead of defining the vectors of frequencies f1 and f2 as in the previous paragraph, it is possible to import data from a csv file. For example the data of the problem are stored in the file Accommodation.csv as follows:

```
CATEGORIES;f1;f2
Camping;1;4
Hotel;45;97
Bed & Breakfast;2;10
Farm;2;4
House/Flat;4;26
```

In the first column, named CATEGORIES, the modalities of the response variable are listed; the second and third columns contain the observed frequencies. A semicolon is the symbol used to separate columns. By using the command setwd() to specify the path with the working folder, it is then possible to define the vectors of frequencies as follows:

```
> Z = read.csv("Accommodation.csv",header=TRUE,sep = ";", quote="",
+   dec=".",fill = TRUE, comment.char="")
> f1=Z$f1
> f2=Z$f2
```

For a preliminary descriptive analysis of data it could be interesting to use Pareto diagrams. The package qcc should be installed and for each sample with the command pareto.chart() the Pareto diagram can be produced:

```
> pareto.chart(f1)
> pareto.chart(f2)
```

The result is shown in Figure 7.1, where the two graphs are compared. It is worth noting that the prevalent choice falls into category B (Hotel) for both groups, but

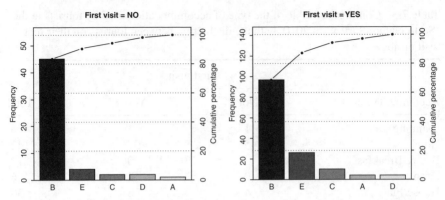

*Figure 7.1   Pareto diagrams of the type of accommodation used by tourists in the 'Sesto Nature Survey' in 2010, by first visit or not. A = Camping; B = Hotel; C = Bed & Breakfast; D = Farm; E = House/Flat.*

the order of the categories in the diagram (according to frequencies) is not the same. The concentration of frequencies in $f_1$ is greater than in $f_2$, thus the heterogeneity of the accommodation in the first sample is lower. The previous commands give the following outputs, representing the distributions of the transformed variables, respectively:

```
Pareto chart analysis for f1
 Frequency   Cum.Freq.   Percentage Cum.Percent.
 B 45.000000 45.000000   83.333333   83.333333
 E 4.000000  49.000000    7.407407   90.740741
 C 2.000000  51.000000    3.703704   94.444444
 D 2.000000  53.000000    3.703704   98.148148
 A 1.000000  54.000000    1.851852  100.000000

Pareto chart analysis for f2
    Frequency  Cum.Freq.   Percentage   Cum.Percent.
 B  97.000000   97.000000  68.794326    68.794326
 E  26.000000  123.000000  18.439716    87.234043
 C  10.000000  133.000000   7.092199    94.326241
 A   4.000000  137.000000   2.836879    97.163121
 D   4.000000  141.000000   2.836879   100.000000
```

The greater level of heterogeneity of sample 2, shown by the reported graphs and the tables, can be proved by computing a sampling index, for example the Shannon's entropy:

```
> S.index.1=het(f1,ind="S")
> S.index.1
[1] 0.6627326
```

```
> S.index.2=het(f2,ind="S")
> S.index.2
[1] 0.9588756
```

The entropy of the sample of returning tourists (sample 1) is equal to 0.663 while the index computed on the sample of new visitors (sample 2) is equal 0.959. This confirms what we suspected.

From the inferential point of view, let us assume that we wish to test the null hypothesis

$$H_0 : Het(Accommodation_1) = Het(Accommodation_2)$$

against the alternative hypothesis

$$H_1 : Het(Accommodation_1) < Het(Accommodation_2),$$

at the significance level $\alpha = 0.10$. Let us use 5000 permutations for the computation of the $p$-values of the test.

```
> F=cbind(f1,f2)
> het.2sample.test(F,alternative="less",nperm=5000)
```

The following output is provided by the software:

```
#######################################################################
# Two-sample permutation test for heterogeneity comparisons:
# Alternative: less
#
# Observed value of Gini's test statistic: 0.1891101
# P-value of Gini's test: 0.02768892
#
# Observed value of Shannon's test statistic: 0.296143
# P-value of Shannon's test: 0.05967613
#
# Observed value of Renyi-3's test statistic: 0.277892
# P-value of Renyi-3's test: 0.02728908
#
# Observed value of Renyi-infinity's test statistic: 0.1917274
# P-value of Renyi-infinity's test: 0.02169132
#######################################################################
```

Hence all the tests give the same result: at the significance level $\alpha = 0.10$ the null hypothesis is rejected in favor of the alternative. The choice of accommodation by tourists for whom this is not their first visit to Sesto is less diversified. Only the test based on Shannon's index provides a $p$-value slightly greater than 0.05

(weak significance). Considering the other tests, there is strong empirical evidence supporting this conclusion.

## 7.4   Two-sided and multisample test

In Section 7.3 the problem of dominance in heterogeneity, that is the one-sided test for heterogeneity comparisons, has been faced. In several real application problems, when two groups are compared in a two-sample test, the alternative hypothesis is not directional. In this case the goal of the study consists of verifying if there is a significant difference between the populations. Hence if the null hypothesis is not true, then there is a difference between the populations. Since the alternative hypothesis specifies no direction for the difference, we have the so called two-sided test.

Formally, given two populations $P_1$ and $P_2$, let us denote with $Het(P_j)$ the heterogeneity of $P_j$. We are interested to test the hypothesis

$$H_0 : Het(P_1) = Het(P_2)$$

against the alternative

$$H_1 : Het(P_1) \neq Het(P_2).$$

For example, in a customer satisfaction survey where the customers are asked to give a qualitative judgment about a product (e.g., poor, fair, good, excellent) or to express their level of satisfaction about it (e.g., very dissatisfied, moderately dissatisfied, moderately satisfied, very satisfied), the response variable is ordered categorical. Hence, unless the variable is transformed into numerical, by subjectively attributing scores to the ordinal categories representing judgments or satisfaction levels, classical indicators for measuring the central tendency or variability of the responses cannot be computed. For this problem central tendency can be measured with the median and, instead of variability, heterogeneity should be evaluated. It could be of interest to compare two groups of customers in terms of heterogeneity of the judgments, for checking if the representativeness of medians, as indicators representing the global evaluation of the groups, are equal or not. This test could be compared with the two-sample test for variance comparison in the presence of quantitative data.

The procedure of the permutation test for this problem is similar to that of dominance in heterogeneity described in the previous paragraph. For the two-sided problem the test statistic should be based on the absolute value of the difference between the sampling indexes: $T_\eta = |\hat{\eta}_1 - \hat{\eta}_2|$. High values of the test statistic lead to the rejection of the null hypothesis in favor of the alternative. The rest of the procedure follows the same steps already described for the test for dominance in heterogeneity: (1) the observed value of the test statistics $T_\eta^o$ is calculated; (2) $B$ independent permutations of the dataset are performed and for each permutation, a permuted table $\{f^*_{j(k)}; k = 1, \ldots, K; j = 1, 2\}$ is obtained and the corresponding permuted value of the test statistic $T^*_\eta$ is calculated; and (3) finally, the $p$-value $\lambda_\eta = \#(T^*_\eta > T^o_\eta)/B$ can be computed and $H_0$ is rejected in favor of $H_1$ when $\lambda_\eta \leq \alpha$.

The procedure could be generalized to include the multisample case. When the heterogeneities of $C$ populations are compared, with $C > 2$, the hypotheses of the problem are

$$H_0 : Het(P_1) = Het(P_2) = \ldots = Het(P_C)$$

and

$$H_1 : Het(P_j) \neq Het(P_r) \text{ for some } j, r \in \{1, \ldots, C\}.$$

Here the test statistic could be based on the absolute values of the differences between the sampling indexes and the index computed on the pooled dataset. Let us consider $C$ independent samples with i.i.d. observations $\mathbf{X}_j = \{X_{ji}, i = 1, \ldots, n_j; n_j > 2\}$, $j = 1, \ldots, C$. Observed data can be displayed in a $C \times K$ contingency table with absolute frequencies $\{f_{jk} = \sum_{i \leq n_j} \mathbb{I}(X_{ji} = A_k), k = 1, \ldots, K, j = 1, \ldots, C\}$. Marginal frequency of the $k$th column (category) is indicated by $f_{\bullet k} = f_{1k} + \ldots + f_{Ck}, k = 1, \ldots, K$; and marginal frequency of the $j$th row is the sample size $n_j, j = 1, \ldots, C$. The test statistic based on the index $\eta$ is $T_\eta = \sum_{j=1}^C |\hat{\eta}_j - \hat{\eta}_\bullet|$, where $\hat{\eta}_j = \eta(f_{j1}/n_j, \ldots, f_{jK}/n_j)$ is the sampling index for sample $j$ and $\hat{\eta}_\bullet = \eta(f_{\bullet 1}/n, \ldots, f_{\bullet K}/n)$ is the sampling index for the pooled dataset. The null hypothesis is rejected for large values of the test statistic, hence the following steps of the testing procedures are similar to those previously described.

## 7.4.1   Customer satisfaction

Passito wine holds a very important place in the Italian wine market. In 2009 a survey was performed to study the demand for this wine in Veneto, a region in the north-east of Italy and to analyze preferences and consumption habits of the Passito wine drinkers. One of the questions was related to the level of liking to drink Passito wine ('How much do you like drinking Passito wine?'). The interviewed could choose among 7 possible ordered levels of satisfaction, from 1 (=*not at all*) to 7 (=*very much*). The response variable is ordered categorical and the median judgment corresponds to the 3rd ordered category (Table 7.4).

We wish to test the hypothesis that the heterogeneity of responses is different in the two groups at the significance level $\alpha = 0.05$. The null hypothesis is $H_0 : Het(Satisfaction_m) = Het(Satisfaction_f)$ and the alternative $H_1 : Het(Satisfaction_m) \neq Het(Satisfaction_f)$. The observed values of sampling indexes can be computed as usual:

```
> f1=c(13,21,42,48,60,31,22)
> f2=c(15,23,40,26,27,13,5)
```

Table 7.4   Contingency table of the level of liking to drink Passito wine, for males and females, in the Passito survey 2009.

| Satisfaction | Gender | | Total |
| --- | --- | --- | --- |
| | Male | Female | |
| 1= not at all | 13 | 15 | 28 |
| 2 | 21 | 23 | 44 |
| 3 | 42 | 40 | 82 |
| 4 | 48 | 26 | 74 |
| 5 | 60 | 27 | 87 |
| 6 | 31 | 13 | 44 |
| 7= very much | 22 | 5 | 27 |
| Total | 237 | 149 | 386 |

```
> index.G.1=het(f1,ind="G")
> index.G.2=het(f2,ind="G")
> index.S.1=het(f1,ind="S")
> index.S.2=het(f2,ind="S")
> index.R3.1=het(f1,ind="R3")
> index.R3.2=het(f2,ind="R3")
> index.Rinf.1=het(f1,ind="Rinf")
> index.Rinf.2=het(f2,ind="Rinf")
> index.G.1
[1] 0.8268974
> index.G.2
[1] 0.821945
> index.S.1
[1] 1.83854
> index.S.2
[1] 1.813464
> index.R3.1
[1] 1.690734
> index.R3.2
[1] 1.661964
> index.Rinf.1
[1] 1.373716
> index.Rinf.2
[1] 1.315067
```

Table 7.5 shows the values computed with the given commands. All the indexes confirm that the heterogeneity of the evaluations of males are slightly greater than that of females.

Table 7.5    Indexes of heterogeneity for level of liking to drink Passito wine, for males and females, in the Passito survey 2009.

| | Gender | |
|---|---|---|
| Index | Male | Female |
| Gini | 0.827 | 0.822 |
| Shannon | 1.839 | 1.813 |
| Rényi(3) | 1.691 | 1.662 |
| Rényi($\infty$) | 1.374 | 1.315 |

To test the hypotheses previously defined, the function `het.2sample.test` should be used. This time the argument `alternative` should take the value `"two.sided"`. We can consider 10000 permutations for the computation of $p$-values.

```
> F=cbind(f1,f2)
> het.2sample.test(F,alternative="two.sided",nperm=10 000)
```

The following output is provided by $R$:

```
########################################################################
# Two-sample permutation test for heterogeneity comparisons:
# Alternative: two.sided
#
# Observed value of Gini's test statistic: 0.004952441
# P-value of Gini's test: 0.709808
#
# Observed value of Shannon's test statistic: 0.02507638
# P-value of Shannon's test: 0.5919316
#
# Observed value of Renyi-3's test statistic: 0.02877008
# P-value of Renyi-3's test: 0.7689962
#
# Observed value of Renyi-infinity's test statistic: 0.05864873
# P-value of Renyi-infinity's test: 0.7535493
########################################################################
```

All the $p$-values are much greater than 0.05, giving empirical evidence that the hypothesis of equal heterogeneities of the evaluations of males and females should not be rejected.

### 7.4.2    Heterogeneity as a measure of uncertainty

A survey to determine the level of perception of people, smokers and nonsmokers, of the hazards of smoking gave the results shown in Table 7.6. The sample consisted of

Table 7.6   Contingency table of perception of the hazard of smoking, for smokers and nonsmokers (Kvam and Vidakovic, 2007).

| | Group | | |
| --- | --- | --- | --- |
| Hazard | Smokers | Nonsmokers | Total |
| Not dangerous | 9 | 3 | 12 |
| Somewhat dangerous | 14 | 6 | 20 |
| Dangerous | 15 | 16 | 31 |
| Very dangerous | 11 | 26 | 37 |
| Total | 49 | 51 | 100 |

100 people randomly selected (see Kvam and Vidakovic, 2007). The response variable is ordered categorical with $K = 4$ ordered modalities corresponding to increasing levels of hazard. As a matter of fact the larger the heterogeneity of answers, the higher the uncertainty about the awareness of risks. Let us compare the heterogeneities of the answers of the two groups, to determine if the uncertainty of respondents depends on whether or not the person smokes. Just by looking at Table 7.6 it seems that the group of nonsmokers is more homogeneous because a large proportion choose the answer 'Very dangerous'. In the group of smokers the frequencies related to the higher levels of hazard are very similar, denoting a higher level of uncertainty about the awareness of risk. Let us suppose to be not interested in knowing whether the uncertainty of a group is greater than that of the other group but just to know whether the uncertainties are not equal.

A comparison of the uncertainty from a descriptive point of view could be based on the Rényi indexes. Run the following code to compute them in $R$

```
> f1=c(9,14,15,11)
> f2=c(3,6,16,26)
> index.R3.1=het(f1,ind="R3")
> index.R3.2=het(f2,ind="R3")
> index.Rinf.1=het(f1,ind="Rinf")
> index.Rinf.2=het(f2,ind="Rinf")
> index.R3.1
[1] 1.333068
> index.R3.2
[1] 0.9002751
> index.Rinf.1
[1] 1.18377
> index.Rinf.2
[1] 0.6737291
```

Both sampling indexes confirm what we noticed looking at the frequency distributions because $\hat{\eta}_{R_3}$ is equal to 1.333 in the sample of smokers and 0.900 in the sample of

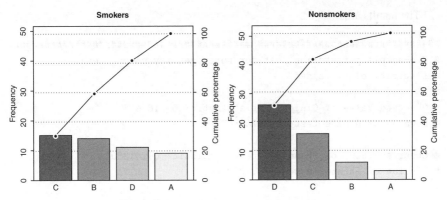

*Figure 7.2    Pareto diagrams of perception of the hazard of smoking, for smokers and nonsmokers. A = Not dangerous; B = Somewhat dangerous; C = Dangerous; D = Very dangerous.*

nonsmokers; $\hat{\eta}_{R\infty}$ is equal to 1.184 for the smokers and 0.674 for the nonsmokers. The Pareto diagrams of the two distributions can be drawn:

```
> pareto.chart(f1)
Pareto chart analysis for f1
  Frequency Cum.Freq. Percentage Cum.Percent.
C 15.00000 15.00000 30.61224  30.61224
B 14.00000 29.00000 28.57143  59.18367
D 11.00000 40.00000 22.44898  81.63265
A  9.00000 49.00000 18.36735 100.00000

> pareto.chart(f2)
Pareto chart analysis for f2
  Frequency  Cum.Freq. Percentage Cum.Percent.
D 26.000000 26.000000 50.980392  50.980392
C 16.000000 42.000000 31.372549  82.352941
B  6.000000 48.000000 11.764706  94.117647
A  3.000000 51.000000 5.882353  100.000000
```

In Figure 7.2 the Pareto diagrams created by the commands above can be compared. The different heterogeneities can be observed comparing both the bar diagrams and the curve of the cumulative frequencies.

Let us test the hypothesis $H_0 : Het(Perception_{smokers}) = Het(Perception_{nonsmokers})$ against $H_1 : Het(Perception_{smokers}) \neq Het(Perception_{nonsmokers})$ at the significance level $\alpha = 0.01$, using the permutation test with 10000 permutations.

```
> F=cbind(f1,f2)
> het.2sample.test(F,alternative="two.sided",nperm=10 000)
```

The results are:

```
##########################################################################
# Two-sample permutation test for heterogeneity comparisons:
# Alternative: two.sided
#
# Observed value of Gini's test statistic: 0.1161495
# P-value of Gini's test: 0.00764847
#
# Observed value of Shannon's test statistic: 0.2413501
# P-value of Shannon's test: 0.00784843
#
# Observed value of Renyi-3's test statistic: 0.4327927
# P-value of Renyi-3's test: 0.00664867
#
# Observed value of Renyi-infinity's test statistic: 0.510041
# P-value of Renyi-infinity's test: 0.00509898
##########################################################################
```

The $p$-values of Rényi-(3)'s test (0.007) and Rényi-($\infty$)'s test (0.005), and also Gini's and Shannon's tests (0.008), are similar and much less than $\alpha$. Hence the null hypothesis of equality in heterogeneity must be rejected in favor of the alternative hypothesis. The heterogeneities of the answers of smokers and nonsmokers are not equal because the awareness (and then uncertainty) about hazards of smoking is different in the two groups.

### 7.4.3    Ethnic heterogeneity

A survey was conducted to test whether ethnicity affects the primary care program choice of students at Tulane University (Doucet *et al.*, 1999). In Table 7.7 the frequency distributions of Medical School applicants according to program (Medicine, Pediatrics or Medicine/Pediatrics) and race (White, Black, Hispanic and Asian) are reported. We wish to test the hypothesis that heterogeneity of ethnicity in the three

Table 7.7    Contingency table of Medical School Applicants broken down by ethnicity and program choice (Doucet *et al.*, 1999).

| Ethnicity | Program | | | Total |
|---|---|---|---|---|
| | Medicine | Pediatrics | Medicine/Pediatrics | |
| White | 30 | 35 | 19 | 84 |
| Black | 11 | 6 | 9 | 26 |
| Hispanic | 3 | 9 | 6 | 18 |
| Asian | 9 | 3 | 8 | 20 |
| Total | 53 | 53 | 42 | 148 |

groups of applicants are different at the significance level $\alpha = 0.05$, with a multisample permutation test with 10000 permutations.

The descriptive analysis based on Gini's and Shannon's indexes leads to the following results:

```
> f1=c(30,11,3,9)
> f2=c(35,6,9,3)
> f3=c(19,9,6,8)
> index.G.1=het(f1,ind="G")
> index.G.2=het(f2,ind="G")
> index.G.3=het(f3,ind="G")
> index.S.1=het(f1,ind="S")
> index.S.2=het(f2,ind="S")
> index.S.3=het(f3,ind="S")
> index.G.1
[1] 0.6044856
> index.G.2
[1] 0.5190459
> index.G.3
[1] 0.6927438
> index.S.1
[1] 1.11211
> index.S.2
[1] 0.9842806
> index.S.3
[1] 1.282778
```

For both indexes the ethnic heterogeneity of applicants in the Medicine/Pediatric program is greater than the heterogeneity of applicants in Medicine which in turn is greater than that of applicants in Pediatrics. The hypotheses of the testing problem are: $H_0 : Het(Ethnicity_{Med}) = Het(Ethnicity_{Ped}) = Het(Ethnicity_{MedPed})$ and $H_1 : H_0$ not true. The commands for performing the permutation test are:

```
> F=cbind(f1,f2,f3)
> het.Csample.test(F,nperm=10 000)
```

The output is the following:

```
############################################################
# Multi-sample permutation test for heterogeneity comparisons:
# Number C of samples: 3
#
# Observed value of Gini's test statistic: 0.183164
# P-value of Gini's test: 0.1252495
#
# Observed value of Shannon's test statistic: 0.3400827
```

```
# P-value of Shannon's test: 0.1082834
#
# Observed value of Renyi-3's test statistic: 0.4896903
# P-value of Renyi-3's test: 0.1192615
#
# Observed value of Renyi-infinity's test statistic: 0.3809858
# P-value of Renyi-infinity's test: 0.1372255
#####################################################################
```

All the tests lead to a very similar result because the *p*-values range from 0.108 to 0.137, hence the null hypothesis of equality in ethnic heterogeneity of the applicants to the three programs should not be rejected at the significance level $\alpha = 0.05$.

### 7.4.4  Reliability analysis

Sometimes in statistical quality control of a production process or a service delivery process the response variable is not numerical. In this case the classical methods such as control charts, capability process, and performance analysis based on statistical methods for numerical data cannot be applied. The parameters of interest will be mode, median, heterogeneity, etc. Let us suppose that a consumer organization, in order to examine the performance of four different brands of batteries for mobile phones, surveys the duration of 60 batteries of *brand 1*, 73 of *brand 2*, 45 of *brand 3* and 55 of *brand 4*. Let us suppose that each battery is classified according to three categories (*Low*, *Medium*, *High*) in reference to the duration. The results are reported in Table 7.8.

Let us perform a multisample permutation test for heterogeneity comparisons in order to test if the performance heterogeneities are equal or not. Let the number of permutations be 5000 and the significance level $\alpha = 0.01$.

The sampling heterogeneities, according to Gini's index, are computed as follows:

```
> f1=c(21,18,21)
> f2=c(12,48,13)
> f3=c(3,32,10)
```

Table 7.8  Contingency table of duration of batteries for mobile phones of four different brands.

| Duration | Brand | | | | |
|---|---|---|---|---|---|
| | 1 | 2 | 3 | 4 | Total |
| Low | 21 | 12 | 3 | 17 | 53 |
| Medium | 18 | 48 | 32 | 15 | 113 |
| High | 21 | 13 | 10 | 23 | 67 |
| Total | 60 | 73 | 45 | 55 | 223 |

```
> f4=c(17,15,23)
> het(f1,ind="G")
[1] 0.665
> het(f2,ind="G")
[1] 0.5089135
> het(f3,ind="G")
[1] 0.4404938
> het(f4,ind="G")
[1] 0.6552066
```

The four indexes seem to be quite different, ranging from 0.440 (*brand 3*) to 0.665 (*brand 1*). Let us compare them using the permutation test:

```
> F=cbind(f1,f2,f3,f4)
> het.Csample.test(F,nperm=5000)
```

The result is as follows:

```
########################################################################
# Multi-sample permutation test for heterogeneity comparisons:
# Number C of samples: 4
#
# Observed value of Gini's test statistic: 0.3707993
# P-value of Gini's test: 9.996002e-05
#
# Observed value of Shannon's test statistic: 0.5409388
# P-value of Shannon's test: 9.996002e-05
#
# Observed value of Renyi-3's test statistic: 1.03189
# P-value of Renyi-3's test: 9.996002e-05
#
# Observed value of Renyi-infinity's test statistic: 1.161476
# P-value of Renyi-infinity's test: 9.996002e-05
########################################################################
```

All the $p$-values are very small. All of them assume values much less than $\alpha$, hence the null hypothesis of equal heterogeneities must be rejected at the significance level $a = 0.01$.

# References

Agresti, A. and Klingenberg, B. (2005) Multivariate tests comparing binomial probabilities, with application to safety studies for drugs. Applied Statistics, 54, 691–706.

Arboretti Giancristofaro, R., Bonnini, S. and Pesarin, F. (2009a) A permutation approach for testing heterogeneity in two-sample problems. Statistics and Computing, 19, 209–216.

Arboretti Giancristofaro, R., Bonnini, S. and Salmaso, L. (2009b) Employment status and education/employment relationship of PhD graduates from the University of Ferrara. Journal of Applied Statistics, 36, 1329–1344.

Brunner, E. and Munzel, U. (2000) The nonparametric Behrens-Fisher problem: asymptotic theory and small-sample approximation. Biometrical Journal, 42, 17–25.

Cohen, A., Kemperman, J., Madigan, D. and Sarkrowitz, H. (2000) Effective directed tests for models with ordered categorical data. Australian and New Zealand Journal of Statistics, 45, 285–300.

Corrain, C., Mezzavilla, F. Pesarin, F. and Scardellato, U. (1977). Il valore discriminativo di alcuni fattori Gm, tra le popolazioni pastorali del Kenya. In: Atti e Memorie dell'Accademia Patavina di Scienze, Lettere ed Arti, LXXXIX, Parte II: Classe di Scienze Matematiche e Naturali. University of Padua, pp. 55–63.

Doucet, H., Shah, M.K., Cummings, T.L. and Kahm, M.J. (1999) Comparison of Internal Medicine, Pediatric and Medicine/Pediatric applicants and factors influencing career choices. Southern Medical Journal, 92, 296–299.

Frosini, B.V. (1981) Heterogeneity indices and distances between distributions. Metron, 34, 95–108.

Gini, C. (1912) Variabilità e Mutabilità. Contributo allo Studio delle Distribuzioni e delle Relazioni Statistiche. Cuppini.

Han, K.E., Catalano, P.J., Senchaudhuri, P. and Mehta, C. (2004) Exact analysis of dose–response for multiple correlated binary outcomes. Biometrics, 4, 216–224.

Hirotsu, C. (1986) Cumulative chi-squared statistic as a tool for testing goodness-of-fit. Biometrika, 73, 165–173.

Kvam, P.H. and Vidakovic, B. (2007) Nonparametric Statistics with Applications to Science and Engineering. John Wiley & Sons, Ltd.

Leti, G. (1965) Sull'entropia, su un indice del Gini e su altre misure dell'eterogeneità di un collettivo. Metron, 24, 332–369.

Loughin, T.M. (2004) A systematic comparison of methods for combining p-values from independent tests. Computational Statistics and Data Analysis, 47, 467–485.

Loughin, T.M. and Scherer, P.N. (1998) Testing for association in contingency tables with multiple column responses. Biometrics, 54, 630–637.

Lumley, T. (1996) Generalized estimating equations for ordinal data: a note on working correlation structures. Biometrics, 52, 354–361.

Nettleton, D. and Banerjee, T. (2001) Testing the equality of distributions of random vectors with categorical components. Computational Statistics and Data Analysis, 37, 195–208.

Patil, G.P. and Taillie, C. (1982) Diversity as a concept and its measurement (with discussion). Journal of the American Statistical Association, 77, 548–567.

Pesarin, F. (1994) Goodness-of-fit testing for ordered discrete distributions by resampling techniques. Metron, 52, 57–71.

Pesarin, F. (2001) Multivariate Permutation Tests: With Application to Biostatistics. John Wiley & Sons, Ltd.

Pesarin, F. and Salmaso, L. (2006) Permutation tests for univariate and multivariate ordered categorical data. Austrian Journal of Statistics, 35, 315–324.

Pesarin, F. and Salmaso, L. (2010) Permutation Tests for Complex Data. Theory, Applications and Software. John Wiley & Sons, Ltd.

Piccolo, D. (2000) Statistica, 2nd edn. Il Mulino.

Pielou, E.C. (1975) Ecological Diversity. John Wiley & Sons, Ltd

Pielou, E.C. (1977) Mathematical Ecology. John Wiley & Sons, Ltd

Rényi, A. (1966) Calculus des probabilitès. Dunod.

Shannon, C.E. (1948) A mathematical theory of communication. Bell System Technological Journal, 27, 379–423.

Shorack, G.R. and Wellner, J.A. (1986) Empirical Processes with Applications to Statistics. John Wiley & Sons, Ltd.

Silvapulle, M.J. and Sen, P.K. (2005) Constrained Statistical Inference, Inequality, Order, and Shape Restrictions. John Wiley & Sons, Ltd.

Troendle, J.F. (2002) A likelihood ratio test for the nonparametric Behrens-Fisher problem. Biometrical Journal, 44, 813–824.

Wang, Y. (1996) A likelihood ratio test against stochastic ordering in several populations. Journal of the American Statistical Association, 91, 1676–1683.

# Appendix A

# Selected critical values for the null distribution of the peak-known Mack–Wolfe statistic

Table A.1 Selected exact critical values for the null distribution of the peak-known Mack–Wolfe $A_p$ statistic : $k$ is the number of treatments, $p$ is the known peak of the umbrella, and $n = n_1 = \cdots = n_k$ is the common sample size.

For given $k$, $n$, $p$, and significance level $\alpha$, the table entry is $a_{p,\alpha}$ satisfying $P_0\{A_p \geq a_{p,\alpha}\} = \alpha$.

| $k = 4$ | $k = 5$ | $k = 6$ |
|---|---|---|
| $p = 3, n = 2$ | $p = 3, n = 2$ | $p = 4, n = 2$ |
| $a_{3,.0893} = 13$ | $a_{3,.0654} = 19$ | $a_{4,.0885} = 26$ |
| $a_{3,.0476} = 14$ | $a_{3,.0377} = 20$ | $a_{4,.0400} = 28$ |
| $a_{3,.0060} = 16$ | $a_{3,.0080} = 22$ | $a_{4,.0075} = 31$ |
| $p = 3, n = 3$ | $p = 3, n = 3$ | $p = 4, n = 3$ |
| $a_{3,.0768} = 27$ | $a_{3,.0882} = 38$ | $a_{4,.0962} = 54$ |
| $a_{3,.0364} = 29$ | $a_{3,.0385} = 41$ | $a_{4,.0419} = 58$ |
| $a_{3,.0080} = 32$ | $a_{3,.0086} = 45$ | $a_{4,.0078} = 64$ |

*(continued)*

*Nonparametric Hypothesis Testing: Rank and Permutation Methods with Applications in R*,
First Edition. Stefano Bonnini, Livio Corain, Marco Marozzi and Luigi Salmaso.
© 2014 John Wiley & Sons, Ltd. Published 2014 by John Wiley & Sons, Ltd.
Companion website: http://www.wiley.com/go/hypothesis_testing

Table A.1   (*Continued*)

| $k = 4$ | $k = 5$ | $k = 6$ |
|---|---|---|
| $p = 3, n = 4$ | $p = 3, n = 4$ | $p = 4, n = 4$ |
| $a_{3,.0836} = 45$ | $a_{3,.0946} = 64$ | $a_{4,.0993} = 92$ |
| $a_{3,.0415} = 48$ | $a_{3,.0475} = 68$ | $a_{4,.0448} = 98$ |
| $a_{3,.0091} = 53$ | $a_{3,.0099} = 75$ | $a_{4,.0094} = 107$ |
| $p = 3, n = 5$ | $p = 3, n = 5$ | $p = 4, n = 5$ |
| $a_{3,.0943} = 67$ | $a_{3,.0944} = 97$ | $a_{4,.0997} = 140$ |
| $a_{3,.0497} = 71$ | $a_{3,.0449} = 103$ | $a_{4,.0469} = 148$ |
| $a_{3,.0094} = 79$ | $a_{3,.0090} = 113$ | $a_{4,.0097} = 161$ |
| | $p = 4, n = 2$ | $p = 5, n = 2$ |
| | $a_{4,.0812} = 21$ | $a_{5,.0837} = 31$ |
| | $a_{4,.0301} = 23$ | $a_{5,.0419} = 33$ |
| | $a_{4,.0077} = 25$ | $a_{5,.0064} = 37$ |
| | $p = 4, n = 3$ | $p = 5, n = 3$ |
| | $a_{4,.0943} = 43$ | $a_{5,.0880} = 65$ |
| | $a_{4,.0452} = 46$ | $a_{5,.0418} = 69$ |
| | $a_{4,.0089} = 51$ | $a_{5,.0077} = 76$ |
| | $p = 4, n = 4$ | $p = 5, n = 4$ |
| | $a_{4,.0973} = 73$ | $a_{5,.0914} = 111$ |
| | $a_{4,.0440} = 78$ | $a_{5,.0447} = 117$ |
| | $a_{4,.0083} = 86$ | $a_{5,.0099} = 127$ |
| | $p = 4, n = 5$ | $p = 5, n = 5$ |
| | $a_{4,.0961} = 111$ | $a_{5,.0931} = 169$ |
| | $a_{4,.0490} = 117$ | $a_{5,.0473} = 177$ |
| | $a_{4,.0086} = 129$ | $a_{5,.0095} = 192$ |

From *Nonparametric Statistical Methods*, 2nd Edition. Reproduced by permission of Wiley. Copyright © 1999, John Wiley & Sons.

# Appendix B

# Selected critical values for the null distribution of the peak-unknown Mack–Wolfe statistic

Table B.1   Selected critical values for the null distribution of the peak-unknown Mack–Wolfe $A_{\hat{p}}^*$ statistic:

$$k = 3, n_1 = n_2 = n_3 = 3(1)10; \quad k = 4(1)10, n_1 = \cdots = n_k = 2(1)10.$$

For given $k$, $n$, and significance level $\alpha$, the table entry is $a_{\hat{p},\alpha}^*$ satisfying $P_0\{A_{\hat{p}}^* \geq a_{\hat{p},\alpha}^*\} \approx \alpha$.

| $k$ | $n = n_1 = \cdots = n_k$ | $a_{\hat{p},.01}^*$ | $a_{\hat{p},.05}^*$ | $a_{\hat{p},.10}^*$ |
|---|---|---|---|---|
| 3 | 3 | 2.556 | 2.324 | 1.889 |
| | 4 | 2.635 | 2.196 | 1.850 |
| | 5 | 2.694 | 2.166 | 1.849 |
| | 6 | 2.668 | 2.102 | 1.787 |
| | 7 | 2.674 | 2.158 | 1.836 |
| | 8 | 2.633 | 2.082 | 1.837 |
| | 9 | 2.623 | 2.111 | 1.800 |
| | 10 | 2.662 | 2.112 | 1.825 |

(*continued*)

*Nonparametric Hypothesis Testing: Rank and Permutation Methods with Applications in R*,
First Edition. Stefano Bonnini, Livio Corain, Marco Marozzi and Luigi Salmaso.
© 2014 John Wiley & Sons, Ltd. Published 2014 by John Wiley & Sons, Ltd.
Companion website: http://www.wiley.com/go/hypothesis_testing

Table B.1    (*Continued*)

| $k$ | $n = n_1 = \cdots = n_k$ | $a^*_{\hat{p},.01}$ | $a^*_{\hat{p},.05}$ | $a^*_{\hat{p},.10}$ |
|---|---|---|---|---|
| 4 | 2 | 2.554 | 2.195 | 1.915 |
|   | 3 | 2.700 | 2.213 | 1.903 |
|   | 4 | 2.708 | 2.180 | 1.912 |
|   | 5 | 2.738 | 2.221 | 1.951 |
|   | 6 | 2.646 | 2.160 | 1.903 |
|   | 7 | 2.744 | 2.205 | 1.898 |
|   | 8 | 2.756 | 2.184 | 1.890 |
|   | 9 | 2.794 | 2.201 | 1.891 |
|   | 10 | 2.771 | 2.172 | 1.876 |
| 5 | 2 | 2.619 | 2.191 | 1.894 |
|   | 3 | 2.725 | 2.239 | 1.963 |
|   | 4 | 2.744 | 2.195 | 1.963 |
|   | 5 | 2.716 | 2.222 | 1.968 |
|   | 6 | 2.749 | 2.227 | 1.972 |
|   | 7 | 2.761 | 2.240 | 1.951 |
|   | 8 | 2.765 | 2.216 | 1.937 |
|   | 9 | 2.786 | 2.236 | 1.925 |
|   | 10 | 2.772 | 2.249 | 1.943 |
| 6 | 2 | 2.643 | 2.226 | 1.964 |
|   | 3 | 2.733 | 2.242 | 2.040 |
|   | 4 | 2.862 | 2.265 | 1.939 |
|   | 5 | 2.851 | 2.251 | 1.969 |
|   | 6 | 2.817 | 2.242 | 1.964 |
|   | 7 | 2.808 | 2.257 | 1.950 |
|   | 8 | 2.819 | 2.256 | 1.963 |
|   | 9 | 2.770 | 2.266 | 1.978 |
|   | 10 | 2.863 | 2.278 | 1.982 |
| 7 | 2 | 2.756 | 2.233 | 1.992 |
|   | 3 | 2.782 | 2.286 | 1.982 |
|   | 4 | 2.802 | 2.279 | 1.999 |
|   | 5 | 2.831 | 2.312 | 2.017 |
|   | 6 | 2.785 | 2.280 | 1.974 |
|   | 7 | 2.823 | 2.294 | 1.997 |
|   | 8 | 2.889 | 2.282 | 1.968 |
|   | 9 | 2.826 | 2.276 | 1.986 |
|   | 10 | 2.919 | 2.338 | 2.006 |

Table B.1    (*Continued*)

| $k$ | $n = n_1 = \cdots = n_k$ | $a^*_{\hat{p},.01}$ | $a^*_{\hat{p},.05}$ | $a^*_{\hat{p},.10}$ |
|---|---|---|---|---|
| 8 | 2 | 2.723 | 2.292 | 2.016 |
| | 3 | 2.821 | 2.297 | 2.021 |
| | 4 | 2.866 | 2.310 | 2.039 |
| | 5 | 2.798 | 2.289 | 2.022 |
| | 6 | 2.885 | 2.339 | 2.027 |
| | 7 | 2.928 | 2.321 | 2.034 |
| | 8 | 2.875 | 2.315 | 2.034 |
| | 9 | 2.874 | 2.305 | 2.021 |
| | 10 | 2.893 | 2.333 | 2.028 |
| 9 | 2 | 2.789 | 2.287 | 1.999 |
| | 3 | 2.815 | 2.283 | 2.027 |
| | 4 | 2.864 | 2.305 | 2.031 |
| | 5 | 2.917 | 2.310 | 2.035 |
| | 6 | 2.887 | 2.325 | 2.041 |
| | 7 | 2.925 | 2.341 | 2.030 |
| | 8 | 2.879 | 2.325 | 2.037 |
| | 9 | 2.883 | 2.293 | 2.027 |
| | 10 | 2.888 | 2.340 | 2.059 |
| 10 | 2 | 2.818 | 2.315 | 2.021 |
| | 3 | 2.802 | 2.315 | 2.026 |
| | 4 | 2.910 | 2.331 | 2.031 |
| | 5 | 2.874 | 2.319 | 2.025 |
| | 6 | 2.912 | 2.297 | 2.027 |
| | 7 | 2.922 | 2.347 | 2.046 |
| | 8 | 2.895 | 2.343 | 2.031 |
| | 9 | 2.948 | 2.380 | 2.050 |
| | 10 | 2.905 | 2.351 | 2.032 |

# Appendix C

# Selected upper-tail probabilities for the null distribution of the Page $L$ statistic

Table C.1   Selected upper-tail probabilities for the null distribution of the Page $L$ statistic:

$$k = 3, n = 2(1)15; k = 4(1)8, n = 2(1)10.$$

For given $k$ and $n$, the table entry for the point $x$ is $P_0\{L \geq x\}$. Under these conditions, if $x$ is such that $P_0\{L \geq x\} = \alpha$, then $\ell_\alpha = x$.

| | | | | | $k = 3$ | | | |
|---|---|---|---|---|---|---|---|---|
| $n$ | $x$ | $P_0\{L \geq x\}$ | $n$ | $x$ | $P_0\{L \geq x\}$ | $n$ | $x$ | $P_0\{L \geq x\}$ |
| 2 | 27 | .1389 | 7 | 89 | .1173 | 10 | 126 | .1112 |
| | 28 | .0278 | | 90 | .0716 | | 127 | .0738 |
| | | | | 91 | .0409 | | 128 | .0466 |
| 3 | 39 | .1528 | | 92 | .0211 | | 129 | .0279 |
| | 40 | .0880 | | 93 | .0097 | | 130 | .0157 |
| | 41 | .0324 | | 94 | .0040 | | 131 | .0083 |
| | 42 | .0046 | | 95 | .0014 | | 132 | .0041 |
| | | | | 96 | .0004 | | 133 | .0018 |

*Nonparametric Hypothesis Testing: Rank and Permutation Methods with Applications in R*,
First Edition. Stefano Bonnini, Livio Corain, Marco Marozzi and Luigi Salmaso.
© 2014 John Wiley & Sons, Ltd. Published 2014 by John Wiley & Sons, Ltd.
Companion website: http://www.wiley.com/go/hypothesis_testing

Table C.1    (*Continued*)

|  |  | $k = 3$ |  |  |  |  |  |  |
|---|---|---|---|---|---|---|---|---|
| $n$ | $x$ | $P_0\{L \geq x\}$ | $n$ | $x$ | $P_0\{L \geq x\}$ | $n$ | $x$ | $P_0\{L \geq x\}$ |
| 4 | 52 | .1088 |  |  |  |  | 134 | .0007 |
|  | 53 | .0563 | 8 | 101 | .1331 |  |  |  |
|  | 54 | .0255 |  | 102 | .0862 | 11 | 141 | .05 |
|  | 55 | .0069 |  | 103 | .0521 |  | 144 | .01 |
|  | 56 | .0008 |  | 104 | .0295 |  | 147 | .001 |
|  |  |  |  | 105 | .0154 |  |  |  |
| 5 | 64 | .1412 |  | 106 | .0073 | 12 | 153 | .05 |
|  | 65 | .0795 |  | 107 | .0031 |  | 156 | .01 |
|  | 66 | .0394 |  | 108 | .0012 |  | 160 | .001 |
|  | 67 | .0181 |  | 109 | .0004 |  |  |  |
|  | 68 | .0066 |  |  |  | 13 | 165 | .05 |
|  | 69 | .0014 | 9 | 113 | .1472 |  | 169 | .01 |
|  | 70 | .0001 |  | 114 | .0990 |  | 172 | .001 |
|  |  |  |  | 115 | .0633 |  |  |  |
| 6 | 76 | .1596 |  | 116 | .0381 | 14 | 178 | .05 |
|  | 77 | .0996 |  | 117 | .0215 |  | 181 | .01 |
|  | 78 | .0572 |  | 118 | .0113 |  | 185 | .001 |
|  | 79 | .0288 |  | 119 | .0055 |  |  |  |
|  | 80 | .0131 |  | 120 | .0024 | 15 | 190 | .05 |
|  | 81 | .0053 |  | 121 | .0009 |  | 194 | .01 |
|  | 82 | .0016 |  |  |  |  | 197 | .001 |
|  | 83 | .0003 |  |  |  |  |  |  |

|  |  | $k = 4$ |  |  |  |  |  |  |
|---|---|---|---|---|---|---|---|---|
| $n$ | $x$ | $P_0\{L \geq x\}$ | $n$ | $x$ | $P_0\{L \geq x\}$ | $n$ | $x$ | $P_0\{L \geq x\}$ |
| 2 | 55 | .1476 | 4 | 108 | .1008 | 5 | 138 | .0253 |
|  | 56 | .1059 |  | 109 | .0724 |  | 139 | .0167 |
|  | 57 | .0556 |  | 110 | .0504 |  | 140 | .0106 |
|  | 58 | .0313 |  | 111 | .0337 |  | 141 | .0065 |
|  | 59 | .0122 |  | 112 | .0217 |  | 142 | .0037 |
|  | 60 | .0017 |  | 113 | .0130 |  | 144 | .0010 |
|  |  |  |  | 114 | .0075 |  | 145 | .0005 |
| 3 | 82 | .1017 |  | 115 | .0039 |  |  |  |
|  | 83 | .0702 |  | 116 | .0017 | 6 | 159 | .1176 |
|  | 84 | .0446 |  | 117 | .0007 |  | 160 | .0916 |
|  | 85 | .0201 |  |  |  |  | 162 | .0524 |
|  | 86 | .0148 | 5 | 133 | .1266 |  | 163 | .0383 |
|  | 87 | .0070 |  | 134 | .0966 |  | 164 | .0273 |
|  | 88 | .0029 |  | 136 | .0524 |  | 165 | .0190 |
|  | 89 | .0007 |  | 137 | .0370 |  | 166 | .0128 |

(*continued*)

Table C.1    (*Continued*)

| | | | | | $k = 4$ | | | |
|---|---|---|---|---|---|---|---|---|
| $n$ | $x$ | $P_0\{L \geq x\}$ | $n$ | $x$ | $P_0\{L \geq x\}$ | $n$ | $x$ | $P_0\{L \geq x\}$ |
| 6 | 167 | .0084 | 8 | 211 | .1010 | 9 | 245 | .0113 |
| | 168 | .0053 | | 212 | .0807 | | 246 | .0081 |
| | 169 | .0032 | | 213 | .0636 | | 247 | .0057 |
| | 171 | .0010 | | 214 | .0493 | | 248 | .0039 |
| | 172 | .0005 | | 216 | .0283 | | 251 | .0011 |
| | | | | 217 | .0208 | | 252 | .0007 |
| 7 | 185 | .1091 | | 219 | .0107 | | | |
| | 186 | .0862 | | 220 | .0075 | 10 | 262 | .1054 |
| | 188 | .0512 | | 221 | .0051 | | 263 | .0866 |
| | 189 | .0383 | | 222 | .0034 | | 265 | .0565 |
| | 190 | .0282 | | 224 | .0014 | | 266 | .0448 |
| | 191 | .0203 | | 225 | .0008 | | 268 | .0271 |
| | 192 | .0143 | | | | | 269 | .0207 |
| | 193 | .0098 | 9 | 236 | .1145 | | 271 | .0116 |
| | 194 | .0066 | | 237 | .0935 | | 272 | .0085 |
| | 195 | .0043 | | 239 | .0600 | | 273 | .0061 |
| | 197 | .0016 | | 240 | .0471 | | 274 | .0043 |
| | 198 | .0010 | | 242 | .0279 | | 277 | .0014 |
| | | | | 243 | .0209 | | 278 | .0009 |

| | | | | | $k = 5$ | | | |
|---|---|---|---|---|---|---|---|---|
| $n$ | $x$ | $P_0\{L \geq x\}$ | $n$ | $x$ | $P_0\{L \geq x\}$ | $n$ | $x$ | $P_0\{L \geq x\}$ |
| 2 | 99 | .1231 | 4 | 205 | .0055 | 7 | 332 | .1080 |
| | 100 | .0962 | | 206 | .0038 | | 333 | .0945 |
| | 102 | .0545 | | 209 | .0011 | | 337 | .0524 |
| | 103 | .0381 | | 210 | .0007 | | 338 | .0446 |
| | 104 | .0261 | | | | | 341 | .0264 |
| | 105 | .0168 | 5 | 239 | .1165 | | 342 | .0219 |
| | 106 | .0096 | | 240 | .0997 | | 345 | .0120 |
| | 107 | .0047 | | 243 | .0596 | | 346 | .0097 |
| | 108 | .0022 | | 244 | .0493 | | 348 | .0061 |
| | 109 | .0006 | | 247 | .0265 | | 349 | .0048 |
| | | | | 248 | .0211 | | 354 | .0013 |
| 3 | 146 | .1177 | | 250 | .0129 | | 355 | .0009 |
| | 147 | .0960 | | 251 | .0100 | | | |
| | 149 | .0611 | | 253 | .0057 | 8 | 378 | .1096 |
| | 150 | .0476 | | 254 | .0042 | | 379 | .0968 |
| | 152 | .0272 | | 258 | .0011 | | 383 | .0562 |
| | 153 | .0198 | | 259 | .0009 | | 384 | .0485 |
| | 154 | .0141 | | | | | 388 | .0254 |

Table C.1    (*Continued*)

| | | | | | $k = 5$ | | | | |
|---|---|---|---|---|---|---|---|---|
| $n$ | $x$ | $P_0\{L \geq x\}$ | $n$ | $x$ | $P_0\{L \geq x\}$ | $n$ | $x$ | $P_0\{L \geq x\}$ |
| | 155 | .0097 | 6 | 286 | .1050 | | 389 | .0213 |
| | 156 | .0065 | | 287 | .0907 | | 392 | .0122 |
| | 157 | .0041 | | 290 | .0562 | | 393 | .0100 |
| | 159 | .0014 | | 291 | .0473 | | 396 | .0053 |
| | 160 | .0007 | | 294 | .0269 | | 397 | .0042 |
| | | | | 295 | .0219 | | 402 | .0012 |
| 4 | 193 | .1090 | | 298 | .0113 | | 403 | .0009 |
| | 194 | .0912 | | 299 | .0089 | | | |
| | 196 | .0617 | | 301 | .0054 | 9 | 424 | .1102 |
| | 197 | .0499 | | 302 | .0041 | | 425 | .0981 |
| | 199 | .0315 | | 306 | .0012 | | 430 | .0514 |
| | 200 | .0245 | | 307 | .0009 | | 431 | .0446 |
| | 203 | .0105 | | | | | 434 | .0284 |
| | 204 | .0077 | | | | | 435 | .0241 |
| 9 | 440 | .0100 | 10 | 470 | .1101 | 10 | 486 | .0118 |
| | 441 | .0083 | | 471 | .0985 | | 487 | .0099 |
| | 443 | .0056 | | 476 | .0537 | | 490 | .0057 |
| | 444 | .0045 | | 477 | .0470 | | 491 | .0047 |
| | 450 | .0011 | | 481 | .0265 | | 498 | .0010 |
| | 451 | .0009 | | 482 | .0227 | | 499 | .0005 |

| | | | | | $k = 6$ | | | | |
|---|---|---|---|---|---|---|---|---|
| $n$ | $x$ | $P_0\{L \geq x\}$ | $n$ | $x$ | $P_0\{L \geq x\}$ | $n$ | $x$ | $P_0\{L \geq x\}$ |
| 2 | 162 | .1008 | 5 | 390 | .1065 | 8 | 617 | .1002 |
| | 163 | .0848 | | 391 | .0963 | | 618 | .0924 |
| | 165 | .0589 | | 396 | .0553 | | 624 | .0547 |
| | 166 | .0480 | | 397 | .0492 | | 625 | .0498 |
| | 168 | .0310 | | 402 | .0254 | | 631 | .0271 |
| | 169 | .0241 | | 403 | .0220 | | 632 | .0243 |
| | 172 | .0101 | | 408 | .0101 | | 639 | .0106 |
| | 173 | .0071 | | 409 | .0085 | | 640 | .0094 |
| | 174 | .0049 | | 411 | .0060 | | 644 | .0055 |
| | 177 | .0011 | | 412 | .0050 | | 645 | .0048 |
| | 178 | .0006 | | 419 | .0012 | | 655 | .0010 |
| | | | | 420 | .0009 | | 656 | .0008 |
| 3 | 238 | .1087 | | | | | | |
| | 239 | .0953 | 6 | 466 | .1023 | 9 | 692 | .1018 |
| | 243 | .0531 | | 467 | .0932 | | 693 | .0944 |
| | 244 | .0451 | | 473 | .0505 | | 700 | .0530 |
| | 247 | .0265 | | 474 | .0451 | | 701 | .0485 |

(*continued*)

Table C.1    (*Continued*)

| | | | $k = 6$ | | | | |
|---|---|---|---|---|---|---|---|

| $n$ | $x$ | $P_0\{L \geq x\}$ | $n$ | $x$ | $P_0\{L \geq x\}$ | $n$ | $x$ | $P_0\{L \geq x\}$ |
|---|---|---|---|---|---|---|---|---|
| | 248 | .0218 | | 478 | .0280 | | 707 | .0273 |
| | 251 | .0115 | | 479 | .0247 | | 708 | .0247 |
| | 252 | .0092 | | 485 | .0108 | | 716 | .0102 |
| | 254 | .0055 | | 486 | .0093 | | 717 | .0090 |
| | 255 | .0042 | | 489 | .0058 | | 721 | .0055 |
| | 259 | .0012 | | 490 | .0049 | | 722 | .0048 |
| | 260 | .0008 | | 498 | .0011 | | 732 | .0011 |
| | | | | 499 | .0009 | | 733 | .0010 |
| 4 | 314 | .1093 | | | | | | |
| | 315 | .0976 | 7 | 541 | .1062 | 10 | 767 | .1026 |
| | 320 | .0522 | | 542 | .0975 | | 768 | .0955 |
| | 321 | .0454 | | 549 | .0505 | | 776 | .0511 |
| | 324 | .0291 | | 550 | .0456 | | 777 | .0469 |
| | 325 | .0248 | | 555 | .0263 | | 783 | .0272 |
| | 330 | .0103 | | 556 | .0233 | | 784 | .0247 |
| | 331 | .0085 | | 562 | .0109 | | 792 | .0107 |
| | 333 | .0056 | | 563 | .0095 | | 793 | .0096 |
| | 334 | .0045 | | 567 | .0053 | | 798 | .0053 |
| | 340 | .0010 | | 568 | .0046 | | 799 | .0047 |
| | 341 | .0008 | | 576 | .0012 | | 810 | .0011 |
| | | | | 577 | .0010 | | 811 | .0009 |

| | | | $k = 7$ | | | | |
|---|---|---|---|---|---|---|---|

| $n$ | $x$ | $P_0\{L \geq x\}$ | $n$ | $x$ | $P_0\{L \geq x\}$ | $n$ | $x$ | $P_0\{L \geq x\}$ |
|---|---|---|---|---|---|---|---|---|
| 2 | 245 | .1074 | 5 | 593 | .1036 | 8 | 938 | .1007 |
| | 246 | .0961 | | 594 | .0966 | | 939 | .0953 |
| | 251 | .0515 | | 602 | .0526 | | 949 | .0524 |
| | 252 | .0448 | | 603 | .0484 | | 950 | .0491 |
| | 255 | .0283 | | 610 | .0259 | | 959 | .0263 |
| | 256 | .0241 | | 611 | .0235 | | 960 | .0244 |
| | 260 | .0115 | | 619 | .0101 | | 971 | .0101 |
| | 261 | .0093 | | 620 | .0091 | | 972 | .0092 |
| | 263 | .0059 | | 624 | .0056 | | 978 | .0053 |
| | 264 | .0046 | | 625 | .0050 | | 979 | .0049 |
| | 268 | .0014 | | 636 | .0011 | | 994 | .0010 |
| | 269 | .0010 | | 637 | .0009 | | 995 | .0009 |
| 3 | 362 | .1019 | 6 | 708 | .1039 | 9 | 1052 | .1033 |
| | 363 | .0930 | | 709 | .0976 | | 1053 | .0981 |
| | 369 | .0509 | | 718 | .0524 | | 1064 | .0530 |
| | 370 | .0455 | | 719 | .0486 | | 1065 | .0499 |

Table C.1     (*Continued*)

## $k = 7$

| $n$ | $x$ | $P_0\{L \ge x\}$ | $n$ | $x$ | $P_0\{L \ge x\}$ | $n$ | $x$ | $P_0\{L \ge x\}$ |
|---|---|---|---|---|---|---|---|---|
| | 374 | .0283 | | 727 | .0253 | | 1075 | .0260 |
| | 375 | .0249 | | 728 | .0232 | | 1076 | .0242 |
| | 381 | .0107 | | 736 | .0109 | | 1087 | .0106 |
| | 382 | .0092 | | 737 | .0099 | | 1088 | .0097 |
| | 385 | .0056 | | 743 | .0052 | | 1095 | .0054 |
| | 386 | .0047 | | 744 | .0047 | | 1096 | .0049 |
| | 393 | .0011 | | 756 | .0010 | | 1112 | .0010 |
| | 394 | .0009 | | 757 | .0009 | | 1113 | .0009 |
| 4 | 478 | .1007 | 7 | 823 | .1028 | 10 | 1167 | .1000 |
| | 479 | .0931 | | 824 | .0969 | | 1168 | .0952 |
| | 486 | .0509 | | 834 | .0511 | | 1180 | .0500 |
| | 487 | .0463 | | 835 | .0476 | | 1181 | .0472 |
| | 493 | .0251 | | 843 | .0262 | | 1191 | .0253 |
| | 494 | .0225 | | 844 | .0242 | | 1192 | .0237 |
| | 500 | .0110 | | 854 | .0102 | | 1204 | .0100 |
| | 501 | .0097 | | 855 | .0092 | | 1205 | .0093 |
| | 505 | .0056 | | 861 | .0051 | | 1212 | .0053 |
| | 506 | .0049 | | 862 | .0046 | | 1213 | .0049 |
| | 515 | .0011 | | 875 | .0011 | | 1230 | .0010 |
| | 516 | .0009 | | 876 | .0009 | | 1231 | .0009 |

## $k = 8$

| $n$ | $x$ | $P_0\{L \ge x\}$ | $n$ | $x$ | $P_0\{L \ge x\}$ | $n$ | $x$ | $P_0\{L \ge x\}$ |
|---|---|---|---|---|---|---|---|---|
| 2 | 353 | .1064 | 3 | 522 | .1009 | 4 | 689 | .1030 |
| | 354 | .0983 | | 523 | .0946 | | 690 | .0974 |
| | 361 | .0528 | | 531 | .0534 | | 700 | .0528 |
| | 362 | .0479 | | 532 | .0494 | | 701 | .0494 |
| | 367 | .0281 | | 539 | .0274 | | 710 | .0258 |
| | 368 | .0250 | | 540 | .0250 | | 711 | .0238 |
| | 374 | .0115 | | 549 | .0101 | | 721 | .0102 |
| | 375 | .0100 | | 550 | .0091 | | 722 | .0093 |
| | 379 | .0054 | | 555 | .0050 | | 728 | .0052 |
| | 380 | .0045 | | 556 | .0045 | | 729 | .0047 |
| | 387 | .0011 | | 566 | .0011 | | 742 | .0011 |
| | 388 | .0009 | | 567 | .0009 | | 743 | .0009 |
| 5 | 856 | .1015 | 7 | 1188 | .1025 | 9 | 1519 | .1028 |
| | 857 | .0965 | | 1189 | .0983 | | 1520 | .0991 |
| | 868 | .0530 | | 1203 | .0517 | | 1536 | .0520 |
| | 869 | .0499 | | 1204 | .0491 | | 1537 | .0498 |

(*continued*)

Table C.1   (*Continued*)

| | | | | | $k = 8$ | | | |
|---|---|---|---|---|---|---|---|---|
| $n$ | $x$ | $P_0\{L \geq x\}$ | $n$ | $x$ | $P_0\{L \geq x\}$ | $n$ | $x$ | $P_0\{L \geq x\}$ |
| | 879 | .0264 | | 1216 | .0258 | | 1551 | .0258 |
| | 880 | .0246 | | 1217 | .0244 | | 1552 | .0245 |
| | 892 | .0101 | | 1231 | .0102 | | 1568 | .0103 |
| | 893 | .0093 | | 1232 | .0096 | | 1569 | .0097 |
| | 900 | .0051 | | 1241 | .0051 | | 1579 | .0053 |
| | 901 | .0047 | | 1242 | .0047 | | 1580 | .0050 |
| | 916 | .0011 | | 1261 | .0010 | | 1602 | .0010 |
| | 917 | .0010 | | 1262 | .0009 | | 1603 | .0010 |
| 6 | 1022 | .1028 | 8 | 1354 | .1011 | 10 | 1685 | .1002 |
| | 1023 | .0982 | | 1355 | .0972 | | 1686 | .0967 |
| | 1036 | .0515 | | 1370 | .0510 | | 1703 | .0503 |
| | 1037 | .0488 | | 1371 | .0487 | | 1704 | .0482 |
| | 1048 | .0257 | | 1384 | .0254 | | 1718 | .0258 |
| | 1049 | .0241 | | 1385 | .0240 | | 1719 | .0246 |
| | 1062 | .0100 | | 1400 | .0101 | | 1736 | .0103 |
| | 1063 | .0093 | | 1401 | .0095 | | 1737 | .0098 |
| | 1071 | .0051 | | 1410 | .0053 | | 1748 | .0052 |
| | 1072 | .0047 | | 1411 | .0050 | | 1749 | .0049 |
| | 1089 | .0010 | | 1432 | .0010 | | 1773 | .0010 |
| | 1090 | .0009 | | 1433 | .0010 | | 1774 | .0009 |

From *Nonparametric Statistical Methods*, 2nd Edition. Reproduced by permission of Wiley. Copyright © 1999, John Wiley & Sons.

# Appendix D

# *R* functions and codes

This appendix includes all *R* codes used throughout the book. It is worth noting that we used and fully recognize the validity and authorship of *R* codes existing in *R* packages and produced as complementary material for other books. Among those we refer to the following websites:

- http://static.gest.unipd.it/~salmaso/web/springerbook.htm
- http://homes.stat.unipd.it/livio/?page=software&lang=IT
- http://www.biostat.uni-hannover.de/software.html
- http://web.mit.edu/caughey/www/Site/Code.html
- http://www.wiley.com/go/npc.

## Packages

*Package stats (version 2.15.3)*

- ```
  ansari.test(x, y, alternative = c("two.sided", "less",
  "greater"), exact = NULL, conf.int = FALSE, conf.level
  = 0.95, ... )
  ```
  It performs the Ansari–Bradley two-sample test for a difference in scale parameters.

- ```
  binom.test(x, n, p = 0.5, alternative = c("two.sided", "less",
  "greater"), conf.level = 0.95)
  ```
  It performs an exact test on the proportion/probability of success in a Bernoulli experiment.

*Nonparametric Hypothesis Testing: Rank and Permutation Methods with Applications in R*,
First Edition. Stefano Bonnini, Livio Corain, Marco Marozzi and Luigi Salmaso.
© 2014 John Wiley & Sons, Ltd. Published 2014 by John Wiley & Sons, Ltd.
Companion website: http://www.wiley.com/go/hypothesis_testing

- `chisq.test(x, y = NULL, correct = TRUE, p = rep(1/length(x), length(x)), rescale.p = FALSE,simulate.p.value = FALSE, B = 2000)`
  It performs the chi-square *goodness-of-fit* tests for comparisons of proportions/ probabilities and for independence of categorical variables.

- `cor(x, y = NULL, use = "everything", method = c("pearson", "kendall", "spearman"))`
  It computes the correlation of $x$ and $y$ if these are vectors. If $x$ and $y$ are matrices then the covariances (or correlations) between the columns of $x$ and the columns of $y$ are computed.

- `cor.test(x, y, alternative = c("two.sided", "less", "greater"), method = c("pearson", "kendall", "spearman"), exact = NULL, conf.level = 0.95, continuity = FALSE, ...)`
  It performs the test for association between paired samples, using Pearson's product moment correlation coefficient, Kendall's tau, or Spearman's rho.

- `fisher.test(x, y = NULL, workspace = 200000, hybrid = FALSE, control = list(), or = 1, alternative = "two.sided", conf.int = TRUE, conf.level = 0.95, simulate.p.value = FALSE, B = 2000)`
  It performs the Fisher exact test for the difference between two proportions/ probabilities.

- `friedman.test(y, ...)`
  It performs the Friedman rank test for unreplicated block design or repeated measures.

- `kruskal.test(x, g, ...)`
  It performs the Kruskal–Wallis rank test for the multisample location problem.

- `ks.test(x, y, ..., alternative = c("two.sided", "less", "greater"), exact = NULL)`
  It performs the one- and two-sample Kolmogorov–Smirnov tests.

- `mcnemar.test(x, y = NULL, correct = TRUE)`
  It performs the the McNemar test for paired data (or bivariate responses) with binary variables.

- `wilcox.test(x, y = NULL, alternative = c("two.sided", "less", "greater"), mu = 0, paired = FALSE, exact = NULL, correct = TRUE, conf.int = FALSE, conf.level = 0.95, ...)`
  It performs the one- and two-sample Wilcoxon tests on vectors of data; the latter is also known as the Mann–Whitney test.

*Package crank*

- `page.trend.test(x)`
  It performs the Page test for ordered alternatives.

*Package CvM2SL1Test*

- `cvmtsl1.pval(cvmtsl1.test(x, y), n1, n2)`
  It applies the L1-version Cramér–von Mises procedure, to test whether two independent samples are drawn from the same population, and computes the *p*-value.

- `cvmtsl1.test(x, y)`
  It computes the test statistic of the L1-version Cramér–von Mises two-sample test.

*Package ICSNP*

- `rank.ctest(X, Y = NULL, mu = NULL, scores = "rank", ... )`
  It performs the one-, two- or *C*-sample multivariate rank test on central tendency. Three different score functions are available.

*Package permute*
It allows `kendall.global()` included in the package *vegan* to be run.

*Package pgirmess*

- `kruskalmc(resp, categ, probs = 0.05, cont=NULL)`
  It performs multiple comparison tests between treatments or treatments versus control related to the Kruskal–Wallis test.

*Package vegan*

- `kendall.global(Y, group, nperm = 999)`
  It computes the coefficient of concordance among several judges (variables, species) and tests its significance through a rank test and a permutation test.

# Other source files

*ad.r*

- `ad(x, label, alt)`
  It computes the Anderson–Darling type test statistic for ordered categorical data.

*ad_perm.r*

- `ad.perm(data, B, alt)`
  It performs the Anderson–Darling type permutation test for ordered categorical data by using the test statistic provided by the function `ad`.

*comb.r*

- `comb(pv, fcomb=c("T", "F", "L"))`
  It applies the nonparametric combination for computing a univariate statistic for multivariate or multiple testing problems.

*combine.r*

- `combine(P, fun=c("Tippett", "Fisher", "Liptak", "Direct", "Max_T"), which=c(2, 3), W=NULL)`
  It applies the nonparametric combination for combining partial tests in complex permutation procedures (e.g., in problems with umbrella alternatives, in multiple comparison procedures, and in tests on moments).

*cucconi_test.r*

- `cucconi(x1, x2, B=10000)`
  It performs the permutation Cucconi two-sample test for comparing locations and scales.

*CSP.r*

- `CSP(y, x, C=1000, exact=FALSE)`
  It performs the permutation test for the two-way ANOVA. By means of constrained synchronized permutations, it computes the $p$-values for the tests on the significance of the interaction and main effects.

*dataperm.r*

- `dataperm(sample="is", dataset, group=c(1), type="synchro", B=1000, seed=FALSE)`
  In order to estimate the null permutation distribution of the test statistics, in complex $C$-sample permutation tests ($C \geq 2$), it performs a random sampling from the space of data permutations conditional on the observed dataset, by means of a Monte Carlo procedure via conditional simulation.

*FWEminP.r*

- `FWEminP(P)`
  It computes the multiplicity adjustment for permutation $p$-values using the minP approach.

*heterogeneity.test.r*

- `het(f, ind=c("G", "S", "R3", "Rinf"))`
  It computes some heterogeneity indexes for categorical data as a function of the vector f of observed absolute frequencies. Four different indexes are available.

- `het.2sample.test(Freq, alternative=c("greater", "less", "two.sided"), nperm)`
  It performs a two-sample permutation test for heterogeneity comparisons.

- `het.Csample.test(Freq, nperm)`
  It performs a $C$-sample ($C > 2$) permutation test for heterogeneity comparisons.

*lepage_test.r*

- `lepage(x1, x2, B=10000)`
  It performs the permutation Lepage two-sample test for comparing locations and scales.

*moments_perm.r*

- `moments.perm(data, B, alt="greater", fun="Fisher")`
  It performs the permutation test on moments for ordered categorical data.

*MW.r*

- `MW(x, y)`
  It computes the Mack–Wolfe statistic for peak-unknown umbrella alternatives.

*obrien_test.r*

- `obrien(x1, x2, alt, B=10000)`
  It performs the permutation O'Brien two-sample test for a difference in scale parameters.

*oneWayPerm.r*

- `one.way.perm(X, Y, B=1000)`
  It performs the permutation test for the one-way ANOVA problem, that is the $C$-sample location testing problem.

*pairwise_comparisons.r*

- `perm.csample.pwc(data, B, fun, alpha, type.perm="synchro")`
  It performs the permutation $C$-sample test for location and the related pairwise comparisons.

*pan_test.r*

- `pan(x1, x2, alt, B=10000)`
  It performs the permutation Pan two-sample test for a difference in scale parameters.

*permsign.r*

- `perm.sign(diff, B=1000, fun=FALSE)`
  It performs the (univariate and multivariate) permutation test for symmetry, the permutation test for paired samples and the multivariate extension of the McNemar test for binary variables.

*perm_2samples.r*

- `perm.2samples(data, alt=c("two.sided", "greater", "less"), B=1000)`
  It performs the two-sample permutation test on central tendency.

*twoWayRs.r*

- `two.way.rs(data, B=1000)`
  It computes the permutation test for related samples useful for location problems and for unreplicated complete block designs.

*t2p.r*

- `t2p(T)`
  It computes the permutation $p$-value, more precisely according to the permutation null distribution of a test statistic it computes the permutation significance level function.

*umbrella.r*

- `umbrella(x, y, B=1000)`
  It performs the permutation test for umbrella alternatives.

*umultiaspect.r*

- `u_multi_aspect(m, stat, alt, maspt, ma_u=FALSE)`
  It computes the multivariate distribution and the corresponding significance levels of the test statistics in multiple tests, according to the permutation distribution of data (provided by the function `dataperm`) and the type of alternatives.

# Index

*Nonparametric Hypothesis Testing: Rank and Permutation Methods with Applications in R*,
First Edition. Stefano Bonnini, Livio Corain, Marco Marozzi and Luigi Salmaso.
© 2014 John Wiley & Sons, Ltd. Published 2014 by John Wiley & Sons, Ltd.
Companion website: http://www.wiley.com/go/hypothesis_testing

# WILEY SERIES IN PROBABILITY AND STATISTICS

ESTABLISHED BY WALTER A. SHEWHART AND SAMUEL S. WILKS

Editors: *David J. Balding, Noel A. C. Cressie, Garrett M. Fitzmaurice, Geof H. Givens, Harvey Goldstein, Geert Molenberghs, David W. Scott, Adrian F. M. Smith, Ruey S. Tsay, Sanford Weisberg*

Editors Emeriti: *J. Stuart Hunter, Iain M. Johnstone, Joseph B. Kadane, Jozef L. Teugels*

The *Wiley Series in Probability and Statistics* is well established and authoritative. It covers many topics of current research interest in both pure and applied statistics and probability theory. Written by leading statisticians and institutions, the titles span both state-of-the-art developments in the field and classical methods.

Reflecting the wide range of current research in statistics, the series encompasses applied, methodological and theoretical statistics, ranging from applications and new techniques made possible by advances in computerized practice to rigorous treatment of theoretical approaches.

This series provides essential and invaluable reading for all statisticians, whether in academia, industry, government, or research.

† ABRAHAM and LEDOLTER · Statistical Methods for Forecasting

AGRESTI · Analysis of Ordinal Categorical Data, *Second Edition*

AGRESTI · An Introduction to Categorical Data Analysis, *Second Edition*

AGRESTI · Categorical Data Analysis, *Third Edition*

ALSTON, MENGERSEN and PETTITT (editors) · Case Studies in Bayesian Statistical Modelling and Analysis

ALTMAN, GILL, and McDONALD · Numerical Issues in Statistical Computing for the Social Scientist

AMARATUNGA and CABRERA · Exploration and Analysis of DNA Microarray and Protein Array Data

AMARATUNGA, CABRERA, and SHKEDY · Exploration and Analysis of DNA Microarray and Other High-Dimensional Data, *Second Edition*

ANDĚL · Mathematics of Chance

ANDERSON · An Introduction to Multivariate Statistical Analysis, *Third Edition*

\* ANDERSON · The Statistical Analysis of Time Series

ANDERSON, AUQUIER, HAUCK, OAKES, VANDAELE, and WEISBERG · Statistical Methods for Comparative Studies

ANDERSON and LOYNES · The Teaching of Practical Statistics

ARMITAGE and DAVID (editors) · Advances in Biometry

ARNOLD, BALAKRISHNAN, and NAGARAJA · Records

\* ARTHANARI and DODGE · Mathematical Programming in Statistics

AUGUSTIN, COOLEN, DE COOMAN and TROFFAES (editors) · Introduction to Imprecise Probabilities

\* BAILEY · The Elements of Stochastic Processes with Applications to the Natural Sciences

---

\*Now available in a lower priced paperback edition in the Wiley Classics Library.
†Now available in a lower priced paperback edition in the Wiley–Interscience Paperback Series.

*Now available in a lower priced paperback edition in the Wiley Classics Library.
†Now available in a lower priced paperback edition in the Wiley–Interscience Paperback Series.

---

\*Now available in a lower priced paperback edition in the Wiley Classics Library.
†Now available in a lower priced paperback edition in the Wiley–Interscience Paperback Series.

*Now available in a lower priced paperback edition in the Wiley Classics Library.
†Now available in a lower priced paperback edition in the Wiley–Interscience Paperback Series.

\* HAHN and SHAPIRO · Statistical Models in Engineering

HAHN and MEEKER · Statistical Intervals: A Guide for Practitioners

HALD · A History of Probability and Statistics and their Applications Before 1750

† HAMPEL · Robust Statistics: The Approach Based on Influence Functions

HARTUNG, KNAPP, and SINHA · Statistical Meta-Analysis with Applications

HEIBERGER · Computation for the Analysis of Designed Experiments

HEDAYAT and SINHA · Design and Inference in Finite Population Sampling

HEDEKER and GIBBONS · Longitudinal Data Analysis

HELLER · MACSYMA for Statisticians

HERITIER, CANTONI, COPT, and VICTORIA-FESER · Robust Methods in Biostatistics

HINKELMANN and KEMPTHORNE · Design and Analysis of Experiments, Volume 1: Introduction to Experimental Design, *Second Edition*

HINKELMANN and KEMPTHORNE · Design and Analysis of Experiments, Volume 2: Advanced Experimental Design

HINKELMANN (editor) · Design and Analysis of Experiments, Volume 3: Special Designs and Applications

HOAGLIN, MOSTELLER, and TUKEY · Fundamentals of Exploratory Analysis of Variance

\* HOAGLIN, MOSTELLER, and TUKEY · Exploring Data Tables, Trends and Shapes

\* HOAGLIN, MOSTELLER, and TUKEY · Understanding Robust and Exploratory Data Analysis

HOCHBERG and TAMHANE · Multiple Comparison Procedures

HOCKING · Methods and Applications of Linear Models: Regression and the Analysis of Variance, *Third Edition*

HOEL · Introduction to Mathematical Statistics, *Fifth Edition*

HOGG and KLUGMAN · Loss Distributions

HOLLANDER, WOLFE, and CHICKEN · Nonparametric Statistical Methods, *Third Edition*

HOSMER and LEMESHOW · Applied Logistic Regression, *Second Edition*

HOSMER, LEMESHOW, and MAY · Applied Survival Analysis: Regression Modeling of Time-to-Event Data, *Second Edition*

HUBER · Data Analysis: What Can Be Learned From the Past 50 Years

HUBER · Robust Statistics

† HUBER and RONCHETTI · Robust Statistics, *Second Edition*

HUBERTY · Applied Discriminant Analysis, *Second Edition*

HUBERTY and OLEJNIK · Applied MANOVA and Discriminant Analysis, *Second Edition*

HUITEMA · The Analysis of Covariance and Alternatives: Statistical Methods for Experiments, Quasi-Experiments, and Single-Case Studies, *Second Edition*

HUNT and KENNEDY · Financial Derivatives in Theory and Practice, *Revised Edition*

\*Now available in a lower priced paperback edition in the Wiley Classics Library.
†Now available in a lower priced paperback edition in the Wiley–Interscience Paperback Series.

*Now available in a lower priced paperback edition in the Wiley Classics Library.
†Now available in a lower priced paperback edition in the Wiley–Interscience Paperback Series.

*Now available in a lower priced paperback edition in the Wiley Classics Library.
†Now available in a lower priced paperback edition in the Wiley–Interscience Paperback Series.

LITTLE and RUBIN · Statistical Analysis with Missing Data, *Second Edition*

LLOYD · The Statistical Analysis of Categorical Data

LOWEN and TEICH · Fractal-Based Point Processes

MAGNUS and NEUDECKER · Matrix Differential Calculus with Applications in Statistics and Econometrics, *Revised Edition*

MALLER and ZHOU · Survival Analysis with Long Term Survivors

MARCHETTE · Random Graphs for Statistical Pattern Recognition

MARDIA and JUPP · Directional Statistics

MARKOVICH · Nonparametric Analysis of Univariate Heavy-Tailed Data: Research and Practice

MARONNA, MARTIN and YOHAI · Robust Statistics: Theory and Methods

MASON, GUNST, and HESS · Statistical Design and Analysis of Experiments with Applications to Engineering and Science, *Second Edition*

McCULLOCH, SEARLE, and NEUHAUS · Generalized, Linear, and Mixed Models, *Second Edition*

McFADDEN · Management of Data in Clinical Trials, *Second Edition*

\* McLACHLAN · Discriminant Analysis and Statistical Pattern Recognition

McLACHLAN, DO, and AMBROISE · Analyzing Microarray Gene Expression Data

McLACHLAN and KRISHNAN · The EM Algorithm and Extensions, *Second Edition*

McLACHLAN and PEEL · Finite Mixture Models

McNEIL · Epidemiological Research Methods

MEEKER and ESCOBAR · Statistical Methods for Reliability Data

MEERSCHAERT and SCHEFFLER · Limit Distributions for Sums of Independent Random Vectors: Heavy Tails in Theory and Practice

MENGERSEN, ROBERT, and TITTERINGTON · Mixtures: Estimation and Applications

MICKEY, DUNN, and CLARK · Applied Statistics: Analysis of Variance and Regression, *Third Edition*

\* MILLER · Survival Analysis, *Second Edition*

MONTGOMERY, JENNINGS, and KULAHCI · Introduction to Time Series Analysis and Forecasting

MONTGOMERY, PECK, and VINING · Introduction to Linear Regression Analysis, *Fifth Edition*

MORGENTHALER and TUKEY · Configural Polysampling: A Route to Practical Robustness

MUIRHEAD · Aspects of Multivariate Statistical Theory

MULLER and STOYAN · Comparison Methods for Stochastic Models and Risks

MURTHY, XIE, and JIANG · Weibull Models

MYERS, MONTGOMERY, and ANDERSON-COOK · Response Surface Methodology: Process and Product Optimization Using Designed Experiments, *Third Edition*

---

\*Now available in a lower priced paperback edition in the Wiley Classics Library.
†Now available in a lower priced paperback edition in the Wiley–Interscience Paperback Series.

MYERS, MONTGOMERY, VINING, and ROBINSON · Generalized Linear Models. With Applications in Engineering and the Sciences, *Second Edition*

NATVIG · Multistate Systems Reliability Theory With Applications

† NELSON · Accelerated Testing, Statistical Models, Test Plans, and Data Analyses

† NELSON · Applied Life Data Analysis

NEWMAN · Biostatistical Methods in Epidemiology

NG, TAIN, and TANG · Dirichlet Theory: Theory, Methods and Applications

OKABE, BOOTS, SUGIHARA, and CHIU · Spatial Tesselations: Concepts and Applications of Voronoi Diagrams, *Second Edition*

OLIVER and SMITH · Influence Diagrams, Belief Nets and Decision Analysis

PALTA · Quantitative Methods in Population Health: Extensions of Ordinary Regressions

PANJER · Operational Risk: Modeling and Analytics

PANKRATZ · Forecasting with Dynamic Regression Models

PANKRATZ · Forecasting with Univariate Box-Jenkins Models: Concepts and Cases

PARDOUX · Markov Processes and Applications: Algorithms, Networks, Genome and Finance

PARMIGIANI and INOUE · Decision Theory: Principles and Approaches

* PARZEN · Modern Probability Theory and Its Applications

PEÑA, TIAO, and TSAY · A Course in Time Series Analysis

PESARIN and SALMASO · Permutation Tests for Complex Data: Applications and Software

PIANTADOSI · Clinical Trials: A Methodologic Perspective, *Second Edition*

POURAHMADI · Foundations of Time Series Analysis and Prediction Theory

POURAHMADI · High-Dimensional Covariance Estimation

POWELL · Approximate Dynamic Programming: Solving the Curses of Dimensionality, *Second Edition*

POWELL and RYZHOV · Optimal Learning

PRESS · Subjective and Objective Bayesian Statistics, *Second Edition*

PRESS and TANUR · The Subjectivity of Scientists and the Bayesian Approach

PURI, VILAPLANA, and WERTZ · New Perspectives in Theoretical and Applied Statistics

† PUTERMAN · Markov Decision Processes: Discrete Stochastic Dynamic Programming

QIU · Image Processing and Jump Regression Analysis

* RAO · Linear Statistical Inference and Its Applications, *Second Edition*

RAO · Statistical Inference for Fractional Diffusion Processes

RAUSAND and HØYLAND · System Reliability Theory: Models, Statistical Methods, and Applications, *Second Edition*

RAYNER, THAS, and BEST · Smooth Tests of Goodnes of Fit: Using R, *Second Edition*

RENCHER and SCHAALJE · Linear Models in Statistics, *Second Edition*

RENCHER and CHRISTENSEN · Methods of Multivariate Analysis, *Third Edition*

*Now available in a lower priced paperback edition in the Wiley Classics Library.
†Now available in a lower priced paperback edition in the Wiley–Interscience Paperback Series.

RENCHER · Multivariate Statistical Inference with Applications

RIGDON and BASU · Statistical Methods for the Reliability of Repairable Systems

* RIPLEY · Spatial Statistics

* RIPLEY · Stochastic Simulation

ROHATGI and SALEH · An Introduction to Probability and Statistics, *Second Edition*

ROLSKI, SCHMIDLI, SCHMIDT, and TEUGELS · Stochastic Processes for Insurance and Finance

ROSENBERGER and LACHIN · Randomization in Clinical Trials: Theory and Practice

ROSSI, ALLENBY, and McCULLOCH · Bayesian Statistics and Marketing

† ROUSSEEUW and LEROY · Robust Regression and Outlier Detection

ROYSTON and SAUERBREI · Multivariate Model Building: A Pragmatic Approach to Regression Analysis Based on Fractional Polynomials for Modeling Continuous Variables

* RUBIN · Multiple Imputation for Nonresponse in Surveys

RUBINSTEIN and KROESE · Simulation and the Monte Carlo Method, *Second Edition*

RUBINSTEIN and MELAMED · Modern Simulation and Modeling

RUBINSTEIN, RIDDER, and VAISMAN · Fast Sequential Monte Carlo Methods for Counting and Optimization

RYAN · Modern Engineering Statistics

RYAN · Modern Experimental Design

RYAN · Modern Regression Methods, *Second Edition*

RYAN · Sample Size Determination and Power

RYAN · Statistical Methods for Quality Improvement, *Third Edition*

SALEH · Theory of Preliminary Test and Stein-Type Estimation with Applications

SALTELLI, CHAN, and SCOTT (editors) · Sensitivity Analysis

SCHERER · Batch Effects and Noise in Microarray Experiments: Sources and Solutions

* SCHEFFE · The Analysis of Variance

SCHIMEK · Smoothing and Regression: Approaches, Computation, and Application

SCHOTT · Matrix Analysis for Statistics, *Second Edition*

SCHOUTENS · Levy Processes in Finance: Pricing Financial Derivatives

SCOTT · Multivariate Density Estimation: Theory, Practice, and Visualization

* SEARLE · Linear Models

† SEARLE · Linear Models for Unbalanced Data

† SEARLE · Matrix Algebra Useful for Statistics

† SEARLE, CASELLA, and McCULLOCH · Variance Components

SEARLE and WILLETT · Matrix Algebra for Applied Economics

SEBER · A Matrix Handbook For Statisticians

† SEBER · Multivariate Observations

SEBER and LEE · Linear Regression Analysis, *Second Edition*

† SEBER and WILD · Nonlinear Regression

*Now available in a lower priced paperback edition in the Wiley Classics Library.
†Now available in a lower priced paperback edition in the Wiley–Interscience Paperback Series.

---

*Now available in a lower priced paperback edition in the Wiley Classics Library.
†Now available in a lower priced paperback edition in the Wiley–Interscience Paperback Series.

---

\*Now available in a lower priced paperback edition in the Wiley Classics Library.
†Now available in a lower priced paperback edition in the Wiley–Interscience Paperback Series.